T0181483

Genetic and Evolutionary Computation

Series Editors:
David E. Goldberg
John R. Koza

More information about this series at http://www.springer.com/series/7373

Genetic and Evolutionary Computation

Series Editors
David E. Goldberg
John R. Koza

Rick Riolo • W.P. Worzel • Mark Kotanchek
Arthur Kordon

Editors

Genetic Programming Theory and Practice XIII

 Springer

Editors
Rick Riolo
Center for the Study of Complex Systems
University of Michigan
Ann Arbor, MI, USA

W.P. Worzel
Evolution Enterprises
Ann Arbor, MI, USA

Arthur Kordon (Retired)

Mark Kotanchek
Evolved Analytics
Midland, MI, USA

ISSN 1932-0167
Genetic and Evolutionary Computation
ISBN 978-3-319-81706-4   ISBN 978-3-319-34223-8   (eBook)
DOI 10.1007/978-3-319-34223-8

Printed on acid-free paper

This Springer imprint is published by Springer Nature
The registered company is Springer International Publishing AG
The registered company address is: Gewerbestrasse 11, 6330 Cham, Switzerland

Dedication

This book is dedicated to John Henry Holland, whose work and kindness touched his students, his colleagues, his friends, and many others who knew him only from his writings.

PREQUEL

Before the blank–full of fresh
      grain scent and flecked
            like oatmeal woven flat–
canvas, before the blank canvas
      is stretched or strained
            tight as an egg, before then–
                  sketch. It doesn't catch
            commencement: it won't hook
            the scene like a rug,
or strategize too far ahead.
            It isn't chess. It doesn't expect
            the homestretch or the check.
                  Each line braves rejection
of the every, edits restless
      all into a space that's still
the space of least commitment, distilling
                  latitudes in draft.
      It would domesticate the feral
            dusk and stockpile dawn.
      It would be commensurate, but settles
            for less, settles
prairies in its channels. Great plains
                  roar and waterfall, yawn and frost

between the lines.
From hunger, from blank
and black, it models erotic
stopped tornadoes, the high relief
of trees. In advance or retreat, in terraced
dynamics–its bets are hedged–with no dead-
bolt perspective. Its point of view? One
with the twister in vista glide,
and the cricket in the ditch,
with riverrain, and turbines' trace.
Inside the flux of
flesh and trunk and cloudy come,
within the latent
marrow of the egg, the amber
traveling waves is where
its vantage lies.
Entering the tornado's core,
entering the cricket waltzed by storm–
to confiscate the shifting give
and represent the with-
out which.

—Alice Fulton

Foreword

In 2003, Carl Simon asked Rick Riolo and me to organize a workshop on genetic programming (GP). We decided to bring together people interested in the theory of GP with people whose main focus was applying GP to "real-world" problems and seeing what happens. We also included daily keynote speakers who were in general not familiar with GP but who had challenging ideas in the areas of computer science, commercial applications, and biological sciences. It was originally planned as a one-off workshop, but after the first workshop, there was a lot of enthusiasm to continue it, and so the Genetic Programming Theory and Practice (GPTP) workshop became an annual event at the University of Michigan in Ann Arbor. This book is the 13th such book written by the attendees of GPTP. Over the years, we have had an amazing series of participants working in a wide range of fields who have refined and expanded the understanding and application of GP.

It was entirely fitting then that the first keynote speaker at GPTP was John Holland. For those who may not be familiar with John and his work, he is widely credited with being one of the originators of genetic algorithms and was a founder of the Santa Fe Institute, the Center for the Study of Complex Systems at the University of Michigan, and other key research centers focused on interdisciplinary studies. He received what may have been the first PhD in computer science (from the U of M) in 1959, and his work in complexity theory was central to the development of complexity as an area of serious study.

John was a polymath who came of age in the heady times of computer science when everything is not only seemed possible but inevitable. John never lost the enthusiasm of those days and passed it along to his students, shared it with his colleagues, and brought it to GPTP. As the chain of GPTP workshops unrolled, John would stop in occasionally if there was a speaker he wanted to hear or a topic that intrigued him. Though he never worked with GP himself, he had a knack for going to the heart of a problem and suggesting new ideas and questions that opened new vistas for exploration.

Perhaps more importantly, GPTP is infused with the spirit of the Center for the Study of Complex Systems (CSCS) and the BACH group in particular. As such, it is multidisciplinary and mathematically inclined and looks to find grand patterns from simple principles. This is really no surprise as many of the attendees at GPTP have been students or colleagues of John's. I believe that this world view is also the reason for the longevity of the workshop as its focus is not about this or that technique per se but is about the deeper workings of GP and how to manage it in the application to different problems.

At the memorial held for John at the University of Michigan in October of this year, Stephanie Forrest spoke about what it was like to be John's graduate student, and she described his approach to advising as being the practice of "benign neglect." As a student, she often found this difficult but said she had come to appreciate its virtues and had adopted it with her own students.

I believe that GPTP has benefited from the same quality of benign neglect as CSCS has given us time, space, and support to pursue a complex but fascinating subject for over a decade without bothering about how the workshop was structured, who we invited or how, or even if we published the results. This freedom has become one of the hallmarks of GPTP, and every year, the participants comment on how much they enjoy the workshop as a result.

For more on John's amazing career, the reader is encouraged to read the Santa Fe Institute's obituary at http://www.santafe.edu/news/item/in-memoriam-john-holland/ and, more importantly, to read his numerous, seminal papers and books as he was truly one of the leading founders of our discipline.

Ann Arbor, MI, USA   W.P. Worzel
November 2015

Preface

This book is about the Thirteenth Workshop on Genetic Programming Theory and Practice, a workshop held this year from May 14 to 16, 2015, at the University of Michigan under the auspices of the Center for the Study of Complex Systems. The workshop is a forum for theorists and users of genetic programming to come together and share ideas, insights, and observations. It is designed to be speculative in nature by encouraging participants to discuss ideas or results that are not necessarily ready for peer-reviewed publication.

To facilitate these goals, the time allotted for presentations is longer than is typical at most conferences, and there is also more time devoted for discussion. For example, presenters usually have 40 min to present their ideas and take questions, and then, before each break, there is open discussion on the ideas presented in a session. Additionally, at the end of each day, there is a review of the entire day and the ideas and themes that have emerged during the sessions. Looking back at the schedule, in a typical day, there was 240 min of presentation and 55 min of discussion or fully 19 % of the time spent in open discussion.

In addition to the regular sessions, each day starts with a keynote speaker who gets a full hour of presentation and 10 min of Q&A. By design, the keynotes are generally not about genetic programming but come from a related field or an application area that may be fertile ground for GP. This year, the first keynote speaker was Dave Ackley from the University of New Mexico who delivered and addressed the topic titled "A Requiem for Determinism." This provocative presentation argued that from the beginning of modern computing, people such as John von Neumann argued that hardware could not be relied on to work perfectly in all cases—just because of the nature of electronics in that they will fail some number of times. These days, the growth of complexity of software has added to this problem. Modern software depends on the user's ability to reboot the system when things get out of sync or when hardware fail. Ackley argues that the correct response (as foreseen by von Neumann) is to make systems that continue to function even when the system nominally fails. Dave went on to suggest that given that GP takes its cues from nature, we should consider incorporating methods that survive "mistakes" in execution.

The second keynote speaker was Larry Burns, who had been an executive at General Motors and is now a consultant with Google on their autonomous vehicle project. Larry's talk was about the development of autonomous vehicles and the likely arc of adoption of autonomous vehicles, but he went on to discuss the fact that technology cannot be thought of in isolation and in particular that it exists in a cultural context and is co-dependent on the infrastructure. As engineers, we tend to think only of the technology we are developing, but Larry made a strong case for thinking about work in a larger context.

The third keynote was Julian Togelius on "Games Playing Themselves: Challenges and Opportunities for AI Research in Digital Games." Games have been at the center of AI development since the beginning of modern computers. Turing mused on chess-playing computers. Samuel's checker playing system could be argued to be the beginning of neural nets, at least on an engineering level. Deep Thought attracted worldwide attention when it beat Garry Kasparov, the then-reigning world chess champion. Julian posed a number of interesting questions relating to AI, particularly about the human traits of curiosity and what it means to "like" something. He turned the usual dynamic of interaction around by asking the questions whether games could be "curious" about people and later asked whether computers could "like" games or even "like" making good games. It was an interesting reversal on the usual questions about AI work and was an interesting discussion in the context of GP.

While the keynotes at the workshop were provocative and interesting, the chapters in this book are the core of GPTP. The first chapter by Kommenda et al. is titled "Evolving Simple Symbolic Regression Models by Multi-objective Genetic Programming." This interesting chapter revisits the question of evaluating the complexity of GP expressions as part of the fitness measure for evolution. Most previous efforts focused either on the structural complexity of the expression or an expensive calculation of subtrees and their components. This chapter proposes a lightweight semantic metric which lends itself to efficient multi-modal fitness calculations without using input data.

The second chapter, by Elyasaf et al., titled "Learning Heuristics for Mining RNA Sequence-Structure Motifs" explores the difficult problem correlating RNA sequences to biological functionality. This is a critical problem to finding and understanding biological mechanisms derived from specific RNA sequences. The authors use GP to create hyper-heuristics that find cliques within the graphs of RNA. Though the chapter only describes the approach and does not show concrete results, it is a clever approach to a complex problem, and we look forward to seeing results in a future GPTP.

The next chapter, by de Melo and Banzhaf, "Kaizen Programming for Feature Construction for Classification" adopts the Japanese practice of Kaizen (roughly, continuous improvement) to GP in the domain of classification problems. In this case, they use GP to generate new ideas in the Kaizen algorithm where in this case "ideas" mean classifier rules that are recursively improved, removed, or refined. It is an interesting idea that takes advantage of GP's ability to generate novel partial solutions and then refine them using the Kaizen approach.

In chapter "GP As If You Mean It: An Exercise for Mindful Practice" by William Tozier, Bill argues that pathologies of result in GP sometimes inform us as to the nature of the problem we are trying to solve and that our (learned) instinct of changing GP parameters or even mechanisms to produce a "better" result may be misguided. He goes from there to a practice of learning adapted for GP that can improve how we use GP by being mindful of how it behaves as we change single features in the problem. He borrows from Pickering's *Mangle* to create consistent ways to use GP to learn from the problem rather than to adjust the GP until you get a result you expected.

In chapter "nPool: Massively Distributed Simultaneous Evolution and Cross Validation in EC-Star," Hodjat and Shahrzad continue work on EC-Star, a GP system designed to be massively parallel using the Cloud. This chapter focuses on evolving classifiers by using local populations with k-fold cross-validation that is later tested across different segments of the samples. Additionally, they are developing these classifiers using time series data, which adds an additional challenge to the problem by requiring a lag as part of the operator set. It is a challenging project that has elements of standard cross-validation with island populations but where learning is not permitted between islands and testing is done entirely on different islands with different samples. This creates a danger of premature convergence/overfitting since populations only have one set of samples to learn on, but they control this as compensated for by extensive validation using the other islands. While this is clearly an interesting approach with some good results, the authors suggest that more work needs to be done before it's ready for commercial use.

In chapter "Highly Accurate Symbolic Regression with Noisy Training Data", Michael Korns continues his pursuit of improving an almost plug-and-play approach to solving symbolic regression problems that verge on the pathologic from a GP perspective. Here he introduces an improved algorithm and adds noise to the input data and is able to show that he can still produce excellent results for out-of-sample data. He also makes this system available for further testing by other researchers, inviting them to test it on different symbolic regression problems.

The seventh chapter, by Gustafson et al., is titled "Using Genetic Programming for Data Science: Lessons Learned." The authors are well versed in industrial applications of computational systems and survey the strengths and weaknesses of GP in such applications. They identify a number of areas where GP offers significant value to Data Scientists but also observe some of the faults of GP in such a context. For those seeking to make GP a more accessible technology in the "real world," this chapter should be carefully considered.

The eight chapter is a highly speculative effort by Bill Worzel titled "The Evolution of Everything (EvE) and Genetic Programming." This chapter sets out to explore more open-ended uses of GP. In particular, he focuses on the coming impact of the Internet of Things (sometimes called the Internet of Everything) on the computing world and speculates that with a constant stream of real-world data, GP could break the mold of generational limits and could constantly evolve solutions that change as the world changes. The effort proposes combining GP, functional programming, particulate genes, and neural nets and (most speculatively) suggests that if the singularity is reachable, it probably will be evolved rather than autonomously springing into being.

The ninth chapter, titled "Lexicase Selection for Program Synthesis: A Diversity Analysis," by Spector and Helmuth, is an exploration of the hypothesis that lexicase selection improves diversity in a population. Lexicase selection is compared with tournament selection and implicit fitness sharing. Lexicase showed improved error diversity, which suggests improved population diversity, thus supporting the hypothesis and the expected mechanism for lexicase selection.

In the next chapter, "Behavioral Program Synthesis: Insights and Prospects," by Krawiec et al., the authors argued at the workshop that a single-valued fitness function "abuses" program evolution by forcing it to evolve a lump sum of what is often a complex set of samples. Instead, they propose using an interaction matrix as a more useful metric as it gives information on specific tests.They argue that not only is information being "left on the table" with single-valued metrics but that the overall behavioral characteristic of an evolved solution is lost and a great deal of nuance and understanding goes missing. They then go on to propose what they call behavioral synthesis which focuses on the behavior of evolved solutions as the dominant factor in evolution. This paper suggests that we need a more nuanced notion of fitness.

The eleventh chapter, "Using Graph Databases to Explore the Dynamics of Genetic Programming Runs," McPhee et al. continues the search for understanding diversity in GP populations, a long-standing focus for research in the GP community. However, in this case, the authors are more interested in looking for "critical moments in the dynamics of a run." To do this, they use a graph database to manage the data and then query the database to search for these crucial inflection points. They focus on the question of whether lexicase selection is truly better than tournament selection and why this might be. Though a work still in progress, this chapter suggests that this method of analyzing GP populations is a valuable addition to the GP toolset and re-raises some of the issues explored in chapter "GP As If You Meant It: An Exercise for Mindful Practice" by Tozier about looking at the process and not just the outcome and chapter "Behavioral Program Synthesis: Insights and Prospects" about the study of behavioral synthesis suggesting that this is an area where we will see more study in the near future.

The twelfth chapter is titled "Product Choice with Symbolic Regression and Classification," by Truscott and Korns. This is one of the first, if not the first use of GP in market research. Huge amounts of money are spent surveying customers, and this data is used to predict brand popularity. The authors describe a survey of cell phones and the analysis produced using the ARC symbolic regression system adapted to classification. The results show well compared to existing methods and suggest that more work in this field may be productive.

The thirteenth chapter by Silva et al., is titled "Multiclass Classification Through Multidimensional Clustering" and revisits the difficult problem of multiclass classifications using GP. This builds from their earlier work which mapped values into higher-dimensional space during the training phase and then collected samples into the closest cluster in the higher-order space. This chapter extends this idea by adding a pool of groups of possible GP trees and combining them selectively (via evolution) to create an ensemble of high-dimensional mapping functions. In some ways, this suggests a more transparent version of SVM, and the results presented suggest that this extension produces improved results with less overfitting.

The final chapter was written by Stijven et al. and is titled "Prime-Time Symbolic Regression Takes Its Place in the Real World." With over 25 years of experience in applying symbolic regression to real-world problems, the authors make a strong case for GP to take its place in the frontlines of business. They give examples of how symbolic regression can be applied to business forecasting, commercial process optimization, and policy decision making in addition to their previous demonstration of applications in commercial R&D. Because many business applications are proprietary, they give an example of their methodology, which critically includes careful attention to the design of experiment (DOE) in a model of infectious disease epidemics that can inform policy decisions. All told, it is hard to find a group of people who have done more to advance the acceptance of GP in the real world.

Acknowledgments

We would like to thank all of the participants for again making GP Theory and Practice a successful workshop. As always, it produced a lot of high energy and interesting and topical discussions, debates, and speculations. The keynote speakers added a lot of food for thought and raised some interesting questions about GP's place in the world. We would also like the thank our financial supporters for making the continued existence of GP Theory and Practice possible. These include:

- The Center for the Study of Complex Systems (CSCS)
- John Koza, Third Millenium Venture Capital Limited
- Michael Korns and Gilda Cabral
- Jason Moore, Computational Genetics Laboratory at Dartmouth College
- Mark Kotanchek and Evolved Analytics
- Babak Hodjat at Sentient
- Steve Everist and Everist Life Sciences
- Heuristic and Evolutionary Algorithms Laboratory, Upper Austria University of Applied Science
- Kordon Consulting

A number of people made key contributions to running the workshop and assisting the attendees while they were in Ann Arbor. Foremost among them was Linda Wood and Susan Carpenter, who made this GPTP workshop run smoothly with their diligent efforts before, during, and after the workshop itself. After the workshop, many people provided invaluable assistance in producing this book. Special thanks go to Kala Groscurth who did a wonderful job working with the authors, editors, and publishers and providing editorial and other assistance to get the book completed. Jennifer Malat and Melissa Fearon provided invaluable editorial efforts, from the initial plans for the book through its final publication. Thanks also to Springer for helping with various technical publishing issues.

Ann Arbor, MI, USA   Rick Riolo
Ann Arbor, MI, USA   W.P. Worzel
                                                                       Arthur Kordon
Midland, MI, USA   Mark Kotanchek
November 2015

Contents

**Evolving Simple Symbolic Regression Models
by Multi-Objective Genetic Programming** 1
Michael Kommenda, Gabriel Kronberger, Michael Affenzeller,
Stephan M. Winkler, and Bogdan Burlacu

Learning Heuristics for Mining RNA Sequence-Structure Motifs 21
Achiya Elyasaf, Pavel Vaks, Nimrod Milo, Moshe Sipper,
and Michal Ziv-Ukelson

Kaizen Programming for Feature Construction for Classification 39
Vinícius Veloso de Melo and Wolfgang Banzhaf

GP As If You Meant It: An Exercise for Mindful Practice 59
William A. Tozier

**nPool: Massively Distributed Simultaneous Evolution and
Cross-Validation in EC-Star** 79
Babak Hodjat and Hormoz Shahrzad

Highly Accurate Symbolic Regression with Noisy Training Data 91
Michael F. Korns

Using Genetic Programming for Data Science: Lessons Learned 117
Steven Gustafson, Ram Narasimhan, Ravi Palla, and Aisha Yousuf

The Evolution of Everything (EvE) and Genetic Programming 137
W.P. Worzel

Lexicase Selection for Program Synthesis: A Diversity Analysis 151
Thomas Helmuth, Nicholas Freitag McPhee, and Lee Spector

Behavioral Program Synthesis: Insights and Prospects 169
Krzysztof Krawiec, Jerry Swan, and Una-May O'Reilly

**Using Graph Databases to Explore the Dynamics of Genetic
Programming Runs** 185
Nicholas Freitag McPhee, David Donatucci, and Thomas Helmuth

**Predicting Product Choice with Symbolic Regression
and Classification** 203
Philip Truscott and Michael F. Korns

Multiclass Classification Through Multidimensional Clustering 219
Sara Silva, Luis Muñoz, Leonardo Trujillo,
Vijay Ingalalli, Mauro Castelli, and Leonardo Vanneschi

Prime-Time: Symbolic Regression Takes Its Place in the Real World 241
Sean Stijven, Ekaterina Vladislavleva, Arthur Kordon,
Lander Willem, and Mark E. Kotanchek

Index 261

Contributors

Michael Affenzeller is at the Heuristic and Evolutionary Algorithms Laboratory, University of Applied Sciences Upper Austria, Hagenberg, Austria, and Institute for Formal Models and Verification, Johannes Kepler University, Linz, Austria

Wolfgang Banzhaf is at the Department of Computer Science, Memorial University of Newfoundland, St. John's, NL, Canada,

Bogdan Burlacu is at the Heuristic and Evolutionary Algorithms Laboratory, University of Applied Sciences Upper Austria, Hagenberg, Austria

Institute for Formal Models and Verification, Johannes Kepler University, Linz, Austria

Mauro Castelli is at NOVA IMS, Universidade Nova de Lisboa, Lisbon, Portugal

David Donatucci Division of Science and Mathematics, University of Minnesota, Morris, Morris, MN, USA

Achiya Elyasaf is at the Department of Computer Science, Ben-Gurion University, Beer-Sheva, Israel

Steven Gustafson is at Knowledge Discovery Lab, GE Global Research, Niskayuna, NY, USA

Thomas Helmuth is a PhD candidate in the School of Computer Science at the University of Massachusetts, Amherst, MA, USA

Babak Hodjat is chief scientist and cofounder of Genetic Finance, responsible for the core technology behind the world's largest distributed evolutionary system. Babak is an entrepreneur having started a number of Silicon Valley companies as main inventor and technologist. He was also senior director of engineering at Sybase Anywhere from 2004 to 2008, where he led Mobile Solutions Engineering including the AvantGo Platform and the mBusiness Anywhere and Answers Anywhere product suites. Previously, Babak was the cofounder of CTO and board member of Dejima Inc. acquired by Sybase in April 2004. Babak is the primary inventor of Dejima's patented agent-oriented technology applied to intelligent interfaces for mobile and enterprise computing—the technology behind Apple's Siri. Dejima was one of only four private firms enrolled in the DARPA (Defense Advanced Research Projects Agency)-funded Cognitive Assistant that Learns and Observes (CALO) Project, managed by SRI International and one of the largest AI projects ever funded. Babak served as the acting CEO of Dejima for 9 months from October 2000. In his past experience, he led several large computer networking and machine learning projects at Neda, Inc. Babak received his PhD in machine intelligence from Kyushu University, in Fukuoka, Japan

Vijay Ingalalli is at LIRMM, Montpellier, France

Michael Kommenda is at the Heuristic and Evolutionary Algorithms Laboratory, University of Applied Sciences Upper Austria, Hagenberg, Austria

Institute for Formal Models and Verification, Johannes Kepler University, Linz, Austria

Arthur Kordon (retired) is CEO of Kordon Consulting, Fort Lauderdale, FL, USA

Michael F. Korns is chief technology officer at Analytic Research Foundation, Henderson, NV, USA

Mark E. Kotanchek is chief technology officer of Evolved Analytics, a data modeling consulting and systems company

Krzysztof Krawiec is with the Computational Intelligence Group, Institute of Computing Science, Poznan University of Technology, Poznan, Poland

Gabriel Kronberger is at the Heuristic and Evolutionary Algorithms Laboratory, University of Applied Sciences Upper Austria, Hagenberg, Austria

Nicholas Freitag McPhee is at the Division of Science and Mathematics, University of Minnesota, Morris, MN, USA

Vinícius Veloso de Melo is at the Department of Computer Science, Memorial University of Newfoundland, St. John's, NL, Canada, and Institute of Science and Technology, Federal University of São Paulo (UNIFESP), São Paulo, Brazil

Nimrod Milo is at the Department of Computer Science, Ben-Gurion University, Beer-Sheva, Israel

Luis Muñoz is Tree-Lab, Posgrado en Ciencias de la Ingeniería, Instituto Tecnológico de Tijuana, Tijuana B.C., México

Ram Narasimhan is at GE Digital, San Ramon, CA, USA

Una-May O'Reilly is at ALFA, Computer Science and Artificial Intelligence Laboratory, Massachusetts Institute of Technology, Cambridge, MA, USA

Ravi Palla is at GE Global Research, Niskayuna, NY, USA

Hormoz Shahrzad is principal scientist of Genetic Finance LLC, responsible for the core technology of a massively distributed evolutionary system applied to various domains, including stock trading. Hormoz has been active in the artificial life and artificial intelligence field for more than 20 years

Moshe Sipper is a professor of computer science at Ben-Gurion University, Beer-Sheva, Israel

Lee Spector is a professor of computer science at Hampshire College and an adjunct professor of computer science at the University of Massachusetts, Amherst. He received a B.A. in philosophy from Oberlin College in 1984 and a PhD in computer science from the University of Maryland in 1992. He is the editor-in-chief of the journal *Genetic Programming and Evolvable Machines* and a member of the editorial board of *Evolutionary Computation*. He is also a member of the SIGEVO executive committee, and he was named a fellow of the International Society for Genetic and Evolutionary Computation

Sean Stijven is at the University of Antwerp, Department of Mathematics – Computer Sciences, Antwerpen, Belgium

Jerry Swan is at the York Centre for Complex Systems Analysis, Department of Computer Science, University of York, York, UK

William A. Tozier is at Ann Arbor, MI, USA

Leonardo Trujillo is Tree-Lab, Posgrado en Ciencias de la Ingeniería, Instituto Tecnológico de Tijuana, Tijuana B.C., México

Philip Truscott is at Southwest Baptist University, Bolivar, MO, USA

Pavel Vaks is at the Department of Computer Science, Ben-Gurion University, Beer-Sheva, Israel

Leonardo Vanneschi is at NOVA IMS, Universidade Nova de Lisboa, Lisbon, Portugal

Ekaterina Vladislavleva is at Evolved Analytics, a data modeling consulting and systems company

Lander Willem is at the University of Antwerp, Faculty of Medicine and Health Sciences

Stephan M. Winkler is at the Heuristic and Evolutionary Algorithms Laboratory, University of Applied Sciences Upper Austria, Hagenberg, Austria

W.P. Worzel is one of the original organizers of the first GP Theory and Practice workshop along with Rick Riolo. He is an entrepreneur and a consultant, whose fundamental interest is in understanding the evolutionary mechanisms of GP (and nature) in order to create better GP systems and apply them to new problems

Aisha Yousuf is at GE Global Research, Niskayuna, NY, USA

Michal Ziv-Ukelson is at the Department of Computer Science, Ben-Gurion University, Beer-Sheva, Israel

Evolving Simple Symbolic Regression Models by Multi-Objective Genetic Programming

Michael Kommenda, Gabriel Kronberger, Michael Affenzeller, Stephan M. Winkler, and Bogdan Burlacu

Abstract In this chapter we examine how multi-objective genetic programming can be used to perform symbolic regression and compare its performance to single-objective genetic programming. Multi-objective optimization is implemented by using a slightly adapted version of NSGA-II, where the optimization objectives are the model's prediction accuracy and its complexity. As the model complexity is explicitly defined as an objective, the evolved symbolic regression models are simpler and more parsimonious when compared to models generated by a single-objective algorithm. Furthermore, we define a new complexity measure that includes syntactical and semantic information about the model, while still being efficiently computed, and demonstrate its performance on several benchmark problems. As a result of the multi-objective approach the appropriate model length and the functions included in the models are automatically determined without the necessity to specify them a-priori.

Keywords Symbolic regression • Complexity measures • Multi-objective optimization • Genetic programming • NSGA-II

1 Introduction

Symbolic regression is the task of finding mathematical formulas that model the relationship between several independent and one dependent variable. A distinguishing feature of symbolic regression is that no assumption about the model structure needs to be made a-priori, because the algorithm automatically determines

M. Kommenda (✉) • M. Affenzeller • B. Burlacu
Heuristic and Evolutionary Algorithms Laboratory, University of Applied Sciences Upper Austria, Softwarepark 11, 4232 Hagenberg, Austria

Institute for Formal Models and Verification, Johannes Kepler University, Altenberger Straße 69, 4040 Linz, Austria
e-mail: michael.kommenda@fh-hagenberg.at

G. Kronberger • S.M. Winkler
Heuristic and Evolutionary Algorithms Laboratory, University of Applied Sciences Upper Austria, Softwarepark 11, 4232 Hagenberg, Austria

© Springer International Publishing Switzerland 2016
R. Riolo et al. (eds.), *Genetic Programming Theory and Practice XIII*,
Genetic and Evolutionary Computation, DOI 10.1007/978-3-319-34223-8_1

the necessary model structure to describe the data implicitly. Another benefit due to the model being described as a mathematical formula is that these formulas can be easily interpreted, validated, and incorporated in other programs (Affenzeller et al. 2014). However, the interpretability of symbolic regression models is often hampered by overly complex and large formulas, bloating and introns, and the excessive usage of variables. Furthermore, complex models tend to be overfit and memorize the training data, which results in poor prediction performance on unseen data. Hence, simpler models with similar training accuracy are generally preferred to complex ones.

Symbolic regression problems are commonly solved by genetic programming, where the formulas are generated during the optimization process and internally represented as symbolic expression trees. An approach to avoid overly complex formulas and to limit the growth of the symbolic expression trees is to specify static tree size and depth limits (Koza 1992; Poli et al. 2008). Appropriate values of these two parameters, so that the trees can grow large enough to model the data accurately while avoiding unnecessary complexity, cannot be known a-priori and must be adapted to the concrete problem. Other methods of controlling the tree size include dynamic size limits (Silva and Costa 2009), parsimony pressure methods (Luke and Panait 2002; Poli 2010) and controlling the distribution of tree sizes through so-called Tarpeian bloat control (Dignum and Poli 2008).

In this work, we follow another approach for evolving simple symbolic regression models: we change the problem formulation from single-objective to multi-objective (Smits and Kotanchek 2005), where the prediction errors and the model complexities are simultaneously minimized. Hence, no complexity related parameters values such as the maximum size of the evolved trees and the allowed function symbols have to be predefined, because the multi-objective algorithm implicitly optimizes those as well. Furthermore, no additional methods for bloat or size control have to be incorporated in the algorithm to evolve simple and parsimonious models. The result of such a multi-objective algorithm execution is not a single solution, but a complete Pareto set of models of varying complexity and prediction accuracy. The question remains how to measure the complexity of symbolic regression models and what effects the selected complexity measure has on the overall algorithm performance and to which extent the evolved models differ syntactically.

2 Complexity Measures for Symbolic Regression

Several measures for calculating the complexity of symbolic regression models have been previously proposed. The simplest ones are based only on the characteristics of the symbolic expression tree representing the regression model such as the *tree length* (Eq. (1)) or the *total visitation length* (Eq. (2), also denoted as *expressional complexity* by Smits and Kotanchek (2005)). The visitation length (Keijzer and Foster 2007) has the advantage that it not only includes the size of the trees, but also incorporates information about the skewness of the tree and favors balanced trees. Another proposed complexity measure is the number of variable symbols (either the number of unique variables in the expression, or the total number of leaf nodes representing variables, Eq. (3)). A benefit of those complexity measures is that they can be calculated efficiently within a single tree iteration with the use of caching strategies for already calculated subtree lengths and thus the runtime of the optimization algorithm is hardly affected. A drawback of those complexity measures is that semantic information about the regression models is not included and only the tree shape is taken into account.

This drawback is overcome by the *order of nonlinearity* metric defined by Vladislavleva et al. (2009). The order of nonlinearity is recursively calculated by aggregating the complexity of the subtrees (e.g., $comp(a + b) = max(comp(a), comp(b))$) and includes the minimal degree of a Chebyshev polynomial approximating the response of individual subtrees sufficiently well. This gives an accurate and intuitive definition of the complexity of a regression model, but Chebyshev polynomial approximation can be computationally expensive, although simplifications to reduce the computation time have been proposed, and depends on the range and number of the presented data points.

Another interesting complexity measure is the *functional complexity* (Vanneschi et al. 2010), which is based on the curvature of the model's response. It basically expresses how many times the slope of the model's response changes in each dimension and can be calculated in polynomial time with the number of presented data points. However, the functional complexity includes no information about the tree length or shape and on its own does not prefer smaller models as the other complexity measures do, which is desired when performing multi-objective genetic programming for symbolic regression.

Based on the characteristics of the previously suggested complexity measures, we have derived a new complexity measure that provides an intuitive definition, is independent of the actual data points, and can be calculated efficiently. The goal of this measure is to be used in multi-objective genetic programming to steer the algorithm towards simple symbolic regression models and to strengthen its ability to identify the necessary function symbols to build highly accurate models. The complexity measure is recursively defined by assigning a complexity of 1 to constants and 2 to variable symbols and aggregating the complexity values of those leaf nodes according to specified rules (Eq. (4)). Most of the aggregation rules originate from the mathematical semantics of the encountered symbol. Due to its recursive definition the complexity measure can be calculated with a single iteration of the symbolic expression tree without evaluating the model itself. Another reason for defining the complexity measure in that way, has been that while the complexity of $sin(x) = 2^2 = 4$ is still rather small, the complexity increases exponentially with the size of the subtree beneath the symbol. Therefore, the total complexity of a symbolic regression model is heavily dependent on the level in which more complicated function symbols occur and when performing multi-objective optimization these are pushed towards the leaf nodes of a tree.

An alternative definition could be to reduce the complexity values of constants and variables to 0 and 1 respectively, but this would not penalize large models containing lots of constants. As a result the constant symbol could gain prevalence in the trees of the population and the algorithm would primarily build constant expressions of varying size. The same reasoning applies to the $+1$ term (Eq. (4) line 4) in the case of multiplication and division. 1 is the neutral element to multiplication and therefore the algorithm would build deeply nested tree containing lots of multiplications/divisions with constants and the learning abilities of the algorithm are worsened.

Definitions of complexity measures for symbolic regression:

$$\text{Length}(T) = \sum_{s \in_s T} 1 \tag{1}$$

$$\text{VisitationLength}(T) = \sum_{s \in_s T} \text{Length}(s) \tag{2}$$

$$\text{Variables}(T) = \sum_{s \in_s T} \begin{cases} 1 & \text{if } sym(s) = \text{variable} \\ 0 & \text{otherwise} \end{cases} \tag{3}$$

$s \in_s T$ defines the subtree relation and returns all subtrees s of tree T

$sym(s)$ returns the symbol of the root node of tree s

$$\text{Complexity}(n) = \begin{cases} 1 & \text{if } sym(n) = \text{constant} \\ 2 & \text{if } sym(n) = \text{variable} \\ \sum_{c \in_c n} \text{Complexity}(c) & \text{if } sym(n) \in (+, -) \\ \prod_{c \in_c n} \text{Complexity}(c) + 1 & \text{if } sym(n) \in (*, /) \\ \text{Complexity}(n_1)^2 & \text{if } sym(n) = \text{square} \\ \text{Complexity}(n_1)^3 & \text{if } sym(n) = \text{squareroot} \\ 2^{\text{Complexity}(n_1)} & \text{if } sym(n) \in (\sin, \cos, \tan) \\ 2^{\text{Complexity}(n_1)} & \text{if } sym(n) \in (\exp, \log) \end{cases} \tag{4}$$

$c \in_c n$ defines the child relation and returns all direct child nodes c of node n

indexing is used to refer to the i-th child of a node, i.e. n_1 refers to the first child node of node n

$sym(n)$ returns the symbol of node n

3 NSGA-II for Symbolic Regression

Multi-objective symbolic regression has previously been studied by Smits and Kotanchek (2005) and Vladislavleva et al. (2009), where a novel algorithm called *ParetoGP* has been used. ParetoGP optimizes the accuracy of the models (in terms of the Pearson's correlation coefficient R^2), but in addition to the population a separate archive containing the Pareto front of the best identified models (complexity vs. accuracy) is maintained. New individuals are created by breeding members of the Pareto front with the most accurate models of the population and after each generation the Pareto front is updated. Instead of developing a new algorithm for multi-objective genetic programming, we have used a well-studied multi-objective optimization algorithm that has been adapted to the specific needs when performing symbolic regression.

The nondominated sorting genetic algorithm (NSGA) was proposed by Srinivas and Deb (1994) for solving multi-objective optimization problems. However, its runtime complexity for nondominated sorting is rather high, no elitism is included in the original NSGA formulation and additionally, a sharing parameter for maintaining diversity has to be specified. These points of criticism have been addressed and an improved, faster version called NSGA-II (Deb et al. 2002) has been presented. The major extensions to standard genetic algorithms of NSGA-II are the use of a nondomination rank and crowding distance for guiding the selection towards a uniformly spread Pareto-optimal front. Furthermore, elitism is ensured by combining the parent population and the generated offspring and selecting the best individuals of this set until the new population is filled. The published version of the NSGA-II has been reimplemented in C# based on the published source code[1] and has been integrated in HeuristicLab (Wagner 2009).

3.1 Domination of Solutions with Equal Qualities

To use NSGA-II efficiently for solving multi-objective symbolic regression it has to be adapted to the specifics of multi-objective symbolic regression problems. In the original version of the algorithm solutions with exactly equal objective values are treated as nondominated. This poses a problem when solving symbolic regression problems, because a single-node individual (either a constant value or a variable) will always have a constant quality and complexity value. Furthermore, individuals with only one node are the simplest individuals that can be built and are always included in the Pareto front. Within a few generations of the algorithm, those one-node individuals account for a huge portion of the Pareto front and the algorithm is not able to evolve larger or more complex individuals with a better fit to the presented data. Hence, the domination criterion of NSGA-II has been modified in order to treat solutions with equal objective values as dominated by the first individual with those objective values. This has the effect that only the first one-node solution is included in the Pareto front and results in a better algorithm performance.

The effects of the algorithm adaptations with respect to the domination criterion are displayed in Fig. 1, where the minimum, average and maximum symbolic

---

[1]http://www.iitk.ac.in/kangal/codes.shtml

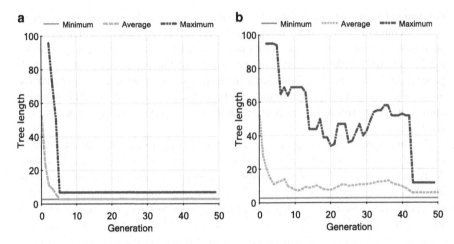

Fig. 1 Comparison of the development of symbolic expression tree length over generations for standard and adapted NSGA-II. The population quickly converges to extremely small trees in the case of the standard implementation of NSGA-II which renders this variant ineffective for symbolic regression. (**a**) Standard NSGA-II. (**b**) Adapted NSGA-II

expression tree length is visualized over generations of the algorithm. On the left side, the behavior of the standard NSGA-II is displayed and it can be seen that the whole population collapses to a few different solutions within the first ten generations. On the right side, the behavior of the adapted NSGA-II is displayed and although the trees get smaller, more diversity is preserved and the algorithm is able to learn from the presented data.

3.2   Discrete Objective Functions

Another aspect when performing symbolic regression is that one of the objective functions describes the fit of the model's output to the presented data, which is in general more important than the simplicity of the models. Frequently, the mean squared error (or a variation thereof) or another correlation criterion such as the Pearson's R^2 correlation are used as an objective function. An issue determined by the floating-point representation of fitness values can arise when many individuals of similar quality (up to many decimal places) and varying complexity artificially enlarge the Pareto front.

A possibility to avoid this issue is to discretize the objective function by rounding the objective value to a fixed number of decimal places. The objective function we used to describe the model accuracy is the Pearson's R^2 correlation of the observed y and the predicted values y'. We round the Pearson's R^2 to three decimal places

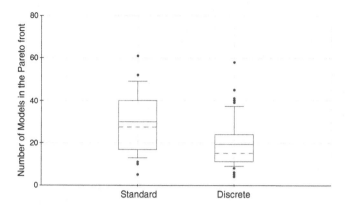

Fig. 2 Number of models in the final Pareto front of 50 repetitions for problem F_1 of NSGA-II with standard and discretized objective functions

resulting in more models having the same prediction accuracy and therefore, a higher selection pressure is applied to build simple models.

Furthermore, the generated Pareto fronts contain fewer models as minor improvements in prediction accuracy are neglected. The differences between a discrete objective function and the standard definition of the Pearson's R^2 are shown in Fig. 2, where the number of models in the final Pareto front of 50 algorithm repetitions on problem F_1 (see Sect. 4.1) are displayed as box plots. By using discrete objective values the size of the Pareto front is almost halved compared to using the exact numeric value.

Illustrative examples of two Pareto fronts extracted from the performed algorithm repetitions are displayed in Fig. 3. The Pareto fronts are shown as the models' normalized mean squared error NMSE (Eq. (5)) and their tree lengths. The NMSE has been used for describing the results, while the Pearson's R^2 is used as an objective during optimization. The reason therefore is that the NMSE is not invariant to translation and scaling (contrary to the R^2) and allows an unbiased comparison on different data partitions such as training and test.

$$\text{NMSE}(y, y') = \frac{1/n \sum (y_i - y_i')^2}{\text{var}(y)} \tag{5}$$

While the Pareto front generated with a discretized objective function (Fig. 3) contains only 11 models, the standard one includes 33 models. The most accurate prediction models have a length of 24 or 91 tree nodes respectively. Another aspect is that the accuracy in the Pareto front without discretized objective values for models larger than 40 nodes increases only by $3.5 \ 10^{-5}$ and can be regarded as irrelevant.

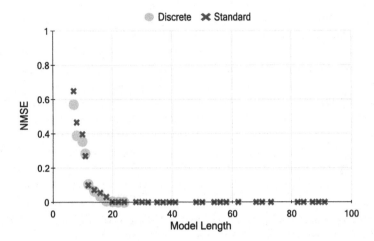

Fig. 3 Exemplary Pareto fronts generate either by an NSGA-II using the standard or the discretized objective function. The Pareto front generated by the discrete objective contains fewer and simpler models that describe the data equally well

4 Experiments

The effectiveness of the new complexity measure is demonstrated by solving five benchmark problems (Sect. 4.1) and compared to single-objective optimization using standard genetic programming as well as using the NSGA-II with other complexity measures. All algorithm variants have been identically configured with the exception that three different maximum tree length values for standard genetic programming and four different objective functions for the NSGA-II (including all previously discussed adaptations) have been tested. Parameters such as the population size, the termination criterion and the allowed terminal or function symbols are listed in Table 1.

The initial population has been created with the probabilistic tree creator (PTC2, Luke 2000), which produces uniformly distributed tree lengths between the specified minimum and maximum length. The individuals for reproduction are selected using a tournament with a group size of four on the prediction accuracy in the case of standard genetic programming, while NSGA-II uses tournament selection with a group size of two on the rank and crowding distance of individuals. A standard subtree swapping crossover, which respects the maximum tree length, has been used as crossover operator and single point, remove branch and replace branch mutation have been applied after crossover with a probability of 25 %. After the reproduction operations the whole previous population gets replaced by the new individuals with the exception of one elite individual when performing standard genetic programming. NSGA-II merges the new and already existing individuals, performs fast non-dominated sorting and keeps the best 1000 individuals which form the new population. The sketched procedure of selection, reproduction and

Table 1 Algorithm settings for the performed experiments (multiple values indicate alternatives)

	Standard GP	NSGA-II
Population size	1000	1000
Maximum generations	500	500
Maximum evaluations	500,000	500,000
Objective function(s)	max R^2	max R^2, min length
		R^2, min visitation length
		R^2, min variables count
		R^2, min complexity
Maximum tree length	20	100
	50	
	100	
Terminal symbols	*constant, weight · variable*	
Function symbols	$+, -, *, /, \sin, \cos, \tan, \exp, \log, x^2, \sqrt{x}$	

replacement is repeated until a specified number of generations are reached. We choose 500 generations as termination which in combination with a population size of 1000 results in 500,000 model evaluations. For every algorithm variant and every problem 50 independent algorithm executions have been conducted to take the stochasticity of the algorithms into account.

4.1 Problems

We have used a wide variety of benchmark problems to test the suitability and the effects of the presented approach. The first experiments were conducted on newly defined benchmark problems (Table 2, Problem $F_1 - F_5$) that have been designed to include polynomial terms and more complex ones containing trigonometric or exponential functions. All input variables x_i were sampled uniformly from $U[-5, 5]$. Due to the fact that these problems do not contain any noise, a model representing the data generating formula can be found and the effects of multi-objective symbolic regression and the new complexity measures can be studied.

In addition, more complex, well known problems, which have been recommended as benchmark symbolic regression problems (White et al. 2013) have been used for testing. The first two problems, Breiman et al. (1984) and Friedman (1991), contain superficial features and have noise added to the dependent variable. The remaining three problems consist of real-world data available at the HeuristicLab website.[2] Hence, these problems cannot be solved exactly and simulate a more practically relevant setting.

---

[2]http://dev.heuristiclab.com/AdditionalMaterial#Real-worlddatasets

Table 2 Description of artificial and real-world problems and the training and test ranges

Name	Function	Training points	Test points
F_1	$F_1(x_1, \ldots, x_5) = x_1 + x_2 + (x_3 - 2)^2 + x_4 x_5$	100	400
F_2	$F_2(x_1, \ldots, x_5) = 10\sin(x_1) + x_2 + x_3 x_4 + x_4 x_5$	100	400
F_3	$F_3(x_1, \ldots, x_5) = \exp(0.7x_1) + (0.5x_2 + 2)^2 + x_3{}^2 + x_4 x_5$	100	400
F_4	$F_4(x_1, \ldots, x_4) = \log((x_1 + x_2)^2)$	100	400
F_5	$F_5(x_1, \ldots, x_4) = (x_1 + x_2)(x_1 + x_2)(x_3 + x_4)$	100	400
Breiman	$F_6(x_1, \ldots, x_{10}) = \begin{cases} 3 + 3x_2 + 2x_3 + x_4 & \text{if } x_1 = 1 \\ -3 + 3x_5 + 2x_6 + x_7 & \text{otherwise} \end{cases}$	5000	5000
Friedman	$F_7(x_1, \ldots, x_{10}) = 0.1e^{4x_1} + 4/[1 + e^{-20x_2 + 10}] + 3x_3 + 2x_4 + x_5 + \epsilon$	5000	5000
Housing	$F_8(x_1, \ldots, x_{13}) = ?$	337	169
Chemical	$F_9(x_1, \ldots, x_{57}) = ?$	711	355
Tower	$F_{10}(x_1, \ldots, x_{25}) = ?$	3136	1863

F_1–F_5 are newly defined problems to demonstrate the effects of different complexity measures, whereas the lower problems have previously been published and used as benchmark problems

4.2 Results

We performed 50 repetitions of each algorithm variant, single-objective with varying maximum tree length and multi-objective with varying complexity measures, on each defined benchmark problem. The results show that multi-objective genetic programming does not worsen the prediction accuracies, while generating simpler models.

Table 3 shows the aggregated information in terms of the median and interquartile range of the prediction accuracy on the training and test partition of the best models obtained in an algorithm run. In the case of multi-objective genetic programming using the NSGA-II the best model is automatically the most complex one (the last model in the Pareto front). Furthermore, the length of the best models is shown to give an indication of their complexity and next to the problem the minimal expression tree length to solve the problem optimally is given.

When comparing the training errors almost all algorithm variants perform equally well, with the exception of standard GP with a length of 100 and the NSGA-II with variable complexity. Among the standard GP algorithms the one with the smallest length constraint performs best, both on the training and test partition. The reason for this is that especially when complex function symbols are allowed to be included in the models, the limitation of the search space helps the algorithm to generate more accurate prediction models, as long as the length constraint is sufficiently high to model the data.

The differences among the NSGA-II algorithms with the new complexity measure, the tree size and the visitation length can be neglected both in terms of the median as well as the interquartile range on the last problems. Only on the first problem the new complexity measure performs better, especially when comparing the interquartile ranges.

Table 3 Performance of the best models of each algorithm variant in terms of the NMSE on the training and test partition and the model length as well as the minimal model length to solve the problem optimally (bold font)

	Training		Test		Length	
	Median	IQR	Median	IQR	Median	IQR
Problem F_1					**18.00**	
GP Length 20	0.001	0.027	0.002	0.031	23.50	1.00
GP Length 50	0.002	0.207	0.002	0.323	51.00	6.00
GP Length 100	0.023	0.209	0.092	0.533	100.50	11.50
NSGA-II Complexity	0.000	0.001	0.001	0.001	27.00	18.50
NSGA-II Visitation Length	0.029	0.246	0.034	0.336	27.00	38.00
NSGA-II Tree Size	0.043	0.199	0.050	0.357	33.00	45.25
NSGA-II Variables	0.165	0.171	0.418	0.504	102.00	8.25
Problem F_2					**18.00**	
GP Length 20	0.000	0.000	0.000	0.000	23.00	1.00
GP Length 50	0.000	0.007	0.000	0.006	52.00	2.00
GP Length 100	0.039	0.418	0.053	0.951	100.00	6.50
NSGA-II Complexity	0.001	0.093	0.001	0.114	32.50	28.25
NSGA-II Visitation Length	0.001	0.001	0.001	0.001	29.50	14.75
NSGA-II Tree Size	0.001	0.001	0.001	0.001	24.00	10.75
NSGA-II Variables	0.001	0.004	0.001	0.006	70.00	45.50
Problem F_3					**20.00**	
GP Length 20	0.005	0.008	0.008	0.016	24.00	1.00
GP Length 50	0.002	0.006	0.006	0.015	52.00	7.75
GP Length 100	0.003	0.101	0.009	0.483	101.50	8.00
NSGA-II Complexity	0.004	0.011	0.006	0.021	31.50	24.50
NSGA-II Visitation Length	0.005	0.019	0.008	0.047	31.00	18.25
NSGA-II Tree Size	0.003	0.011	0.006	0.025	30.50	18.75
NSGA-II Variables	0.051	0.141	0.188	0.557	99.00	27.00
Problem F_4					**11.00**	
GP Length 20	0.000	0.000	0.000	0.052	23.00	2.00
GP Length 50	0.000	0.210	0.042	0.452	51.00	7.00
GP Length 100	0.076	0.354	0.225	0.632	99.00	10.00
NSGA-II Complexity	0.000	0.000	0.010	0.040	11.00	3.00
NSGA-II Visitation Length	0.000	0.000	0.009	0.011	11.00	0.00
NSGA-II Tree Size	0.000	0.000	0.009	0.013	11.00	0.00
NSGA-II Variables	0.000	0.000	0.014	0.014	22.00	19.50
Problem F_5					**12.00**	
GP Length 20	0.025	0.033	0.041	0.045	23.50	2.00
GP Length 50	0.029	0.032	0.046	0.279	52.00	3.00
GP Length 100	0.055	0.112	0.846	8.233	98.00	10.00
NSGA-II Complexity	0.021	0.033	0.042	0.044	22.00	15.25
NSGA-II Visitation Length	0.025	0.033	0.041	0.045	21.00	8.00
NSGA-II Tree Size	0.029	0.033	0.041	0.044	18.50	8.00
NSGA-II Variables	0.034	0.071	0.154	1.065	80.00	60.25

The length of the evolved symbolic regression models for all single-objective genetic programming configurations reaches or slightly exceeds the predefined limit. The length constraint can be exceeded due to the additive and multiplicative linear scaling terms which are added to the models to account for the scaling invariance of the Pearson's R^2. All multi-objective algorithms perform similarly with respect to the model length with the exception of the variable complexity that has almost no selection pressure towards smaller models. Noteworthy is that multi-objective genetic programming finds exactly the data generating formula for the fourth problem F_4 (small predictions errors on the test partition result from slightly inaccurate numerical constants).

Next to the accuracy and length of the final models, we are interested in the functions used in the obtained models. Therefore, we analyzed how often and where trigonometric, exponential and power symbols occur in those models. This is calculated by summing over the size of the affected subtrees whose symbols fall into defined categories (trigonometric: sin, cos, tan—exponential: exp, log—power: $x^2, \sqrt{x}$). If a symbol occurs multiple times all occurrences are counted and the affected subtree size can exceed the model length.

The results of this analysis are displayed in Table 4. The interpretation is eased by comparing the values with the affected subtree size of the shortest model solving the problem exactly (shown next to the problem name). The calculated subtree size can fall below the optimal value for power symbols, because x^2 can be reformulated as $x \cdot x$, yielding a slightly larger model. This happens for example on problem F_1 and F_3.

The standard genetic programming algorithms with a length constraint of 50 and 100 include all available symbols rather often. Standard genetic programming with the smallest length constraint 20 works quite well due to the strict limitation of the search space. NSGA-II with the newly defined complexity measure overall achieves the best results in terms of the affected subtree size of the investigated symbols, which indicates that the combination of syntactical information and the semantics of the symbols, improves the algorithm's ability to determine the necessary complexity to evolve simple yet accurate models. Comparing our complexity measure with the tree size and the visitation length, the last two algorithms generate models with a slightly more complex structure as more nodes are affected by the investigated functions. However, the optimization towards more parsimonious models also helps the algorithm to produce models using fewer trigonometric, exponential or power functions compared to single-objective algorithms using the same length constraints.

4.2.1 Exemplary Models

The advantages of multi-objective symbolic regression are illustrated by the best models (Eqs. (6)–(9)) generated for problem F_2. The best training model out of the 50 repetitions for every algorithm variant has been extracted and after constant folding and numeric optimization all models obtained a test NMSE of at most 10^{-10}.

Table 4 Analysis of the used functions in the best models in terms of the subtree size affected by the symbol grouped into three categories (trigonometric: sin, cos, tan— exponential: exp, log power: x^2, $\sqrt{x}$)

	Trigonometric		Exponential		Power	
	Median	IQR	Median	IQR	Median	IQR
Problem F_1	**0.00**		**0.00**		**4.00**	
GP Length 20	0.00	2.00	0.00	0.00	0.00	4.00
GP Length 50	19.00	31.00	6.00	36.00	8.00	23.00
GP Length 100	56.50	102.00	23.50	93.50	30.00	75.25
NSGA-II Complexity	0.00	0.00	0.00	0.00	2.00	4.00
NSGA-II Visitation Length	2.00	18.75	0.00	4.00	2.00	4.00
NSGA-II Tree Size	0.00	12.00	0.00	8.75	4.00	12.75
NSGA-II Variables	264.00	295.00	144.50	211.50	80.00	126.75
Problem F_2	**2.00**		**0.00**		**0.00**	
GP Length 20	4.00	2.00	0.00	0.00	0.00	0.00
GP Length 50	32.50	40.00	0.00	14.00	0.00	5.75
GP Length 100	150.00	223.75	45.50	129.00	36.50	75.75
NSGA-II Complexity	2.00	9.50	0.00	0.00	0.00	0.00
NSGA-II Visitation Length	6.00	8.75	0.00	0.00	0.00	0.00
NSGA-II Tree Size	6.00	8.00	0.00	0.00	0.00	0.00
NSGA-II Variables	64.50	140.25	22.00	59.25	23.00	67.75
Problem F_3	**0.00**		**2.00**		**6.00**	
GP Length 20	0.00	0.00	4.00	4.00	4.00	2.00
GP Length 50	11.00	21.75	5.50	9.75	9.00	8.75
GP Length 100	84.50	127.25	28.00	52.50	34.00	56.00
NSGA-II Complexity	0.00	0.00	2.00	5.00	5.00	6.25
NSGA-II Visitation Length	0.00	0.00	4.00	4.00	6.00	6.00
NSGA-II Tree Size	0.00	0.00	4.00	4.00	6.00	5.50
NSGA-II Variables	77.00	115.00	28.00	95.00	74.00	127.00
Problem F_4	**0.00**		**5.00**		**4.00**	
GP Length 20	0.00	0.00	17.00	2.00	10.00	17.00
GP Length 50	25.00	32.00	36.00	33.00	21.00	34.50
GP Length 100	144.00	198.00	80.00	120.00	69.00	155.00
NSGA-II Complexity	0.00	0.00	5.00	0.00	4.00	0.00
NSGA-II Visitation Length	0.00	0.00	5.00	0.00	4.00	0.00
NSGA-II Tree Size	0.00	0.00	5.00	0.00	4.00	0.00
NSGA-II Variables	7.00	29.75	16.50	32.00	14.50	25.75
Problem F_5	**0.00**		**0.00**		**4.00**	
GP Length 20	0.00	0.00	0.00	4.00	6.00	8.00
GP Length 50	18.50	32.50	0.00	17.75	12.50	21.00
GP Length 100	72.00	170.00	65.00	83.25	33.50	49.50
NSGA-II Complexity	0.00	0.00	0.00	4.00	0.00	0.00
NSGA-II Visitation Length	0.00	0.00	0.00	0.00	4.00	4.00
NSGA-II Tree Size	0.00	0.00	0.00	4.00	4.00	4.00
NSGA-II Variables	77.50	162.00	62.00	113.25	45.00	82.25

For each problem the minimal subtree size is given for the shortest model solving the problem exactly

Table 5 Size statistics of the best models for Problem-2 per algorithm variant

	Original model		Simplified model		
Problem-2	Length	Depth	Length	Depth	Equation
GP Length 20	18	7	10	4	Eq. (6)
GP Length 50	39	11	20	6	Eq. (8)
GP Length 100	64	21	54	15	Eq. (9)
NSGA-II Complexity	16	7	10	4	Eq. (6)
NSGA-II Visitation Length	14	6	10	4	Eq. (7)
NSGA-II Tree Size	16	7	10	4	Eq. (6)
NSGA-II Variables	25	9	10	4	Eq. (6)

The length and depth of the symbolic expression trees are displayed for their original and simplified version stated in Eqs. (6)–(9)

Therefore, all extracted models explain the relation between the input and output data accurately and there is no difference between the models in terms of prediction quality. GP with a length limit of 20 and NSGA-II with the complexity, tree size and variable measure found exactly the data generating formula f_1 (Eq. (6)), whereas NSGA-II with the visitation length found an alternative formulation f_2 (Eq. (7)). On the contrary, GP with higher length limits of 50 (f_3, Eq. (8)) and 100 (f_4, Eq. (9)) respectively, found models that include additional terms which cannot be removed by constant folding although their impact on the evaluation is minimal.

$$f_1(x) = x_2 + x_3 x_4 + x_4 x_5 + \sin(x_1) \tag{6}$$

$$f_2(x) = x_2 + x_4(x_3 + x_5) + \sin(x_1) \tag{7}$$

$$f_3(x) = x_2 + x_3 x_4 + x_4 x_5 + \sin(x_1)[5.11 \ 10^{-10} x_5/x_1 + 1] \tag{8}$$

$$f_4(x) = x_2 + x_3 x_4 + x_4 x_5 + \sin(x_1) + 8.7 \ 10^{-7} \cos(\cos(\sin(0.99 x_1) + e^{\sin(x_1)})) \tag{9}$$

$$+ 8.85 \ 10^{-7} \cos(\sin(0.79 \sin(x_1) + \sin(e^{\cos(1.61 + e^{\sin(x_1)})})$$

$$\cos(\cos(\sin^2(\cos(\cos(\tan(\sin(0.99 x_1)) + \cos(\sin(0.99 x_1)))))))))$$

The size statistics of the extracted models in their original and simplified version are displayed in Table 5. All models get significantly smaller during the constant folding and simplification operations performed. The models created by GP with a length limit of 20 and NSGA-II found the data generating formula directly (except NSGA-II with the variables complexity measure) and the size reduction during simplification is caused by the transformation of binary trees to n-ary trees. The best model created by NSGA-II variables contained in its original form one additional subtree expressing a constant numerical value that is removed by constant folding. The two GP variants with larger length limits failed to find the data generating formula due to the inclusion of complex subtrees with almost no evaluation effect.

4.2.2 Noisy Data

The same algorithm settings as in the previous experiments have been used for evaluating the performance of the algorithm variants on the five noisy problems. Again 50 repetitions have been performed and the most accurate models with the best performance on the training partition, have been extracted and analyzed. The aggregated information regarding training and test accuracy as well as the model lengths are shown in Table 6.

Contrary to the previously tested artificial problems, GP with a length limit of 20 performs worse compared to the other single-objective algorithms. The reason might be that the smaller length limitation, which gave an advantage on the artificial problems, restricts the search space too much to be able to evolve accurate prediction models.

Due to the noise on the data the training performance can differ significantly the test performance, which is especially apparent on the Housing and Chemical problem. The Breiman, Friedman and Tower problems contain enough data that the effect of the noise is reduced and the difference between the training and test evaluation is minimal. With the exception of the Friedman problem multi-objective symbolic regression with the new complexity measure performs best on all problems. Especially on the Housing and Chemical problems the difference between training and test accuracy is smaller, which might by the preference of less complex functions during model building.

The single-objective algorithms always hit the predefined length limit as it was the case with the results obtained on the artificial problems. The selection pressure towards small models is highest when using the visitation length or tree size as complexity measure for NSGA-II. Hence, these two algorithm variants produced the smallest models, whereas NSGA-II with variable count exhibits no parsimony pressure at all. NSGA-II with the new complexity measure produces models of similar or slightly larger size compared to NSGA-II executions which the size for complexity calculation.

The analysis of the functions in the evolved models, displayed in Table 7, shows a similar picture as the results on the artificial problems. The simplest models, using the fewest trigonometric, exponential and power symbols, have been generated by NSGA-II complexity and GP Length 20 with the difference that the models generated by NSGA-II are more accurate. The largest values in this analysis that indicate more complex models, have been obtained by the other single-objective GP variants and NSGA-II Variables.

5 Conclusion

In this chapter we have investigated the effects of using different complexity measures for multi-objective genetic programming to solve symbolic regression problems and compared the results to standard genetic programming. Multi-

Table 6 Performance of the best models of each algorithm variant in terms of the NMSE on the training and test partition and the model length

	Training		Test		Length	
	Median	IQR	Median	IQR	Median	IQR
Breiman						
GP Length 20	0.263	0.154	0.262	0.158	24.00	1.00
GP Length 50	0.185	0.219	0.185	0.211	53.00	2.75
GP Length 100	0.560	0.430	0.548	0.452	99.50	7.00
NSGA-II Complexity	0.108	0.009	0.109	0.009	70.00	31.25
NSGA-II Visitation Length	0.109	0.017	0.111	0.016	63.00	35.00
NSGA-II Tree Size	0.110	0.013	0.106	0.014	67.00	21.00
NSGA-II Variables	0.134	0.037	0.138	0.038	96.00	20.50
Friedman						
GP Length 20	0.193	0.021	0.190	0.022	24.00	1.00
GP Length 50	0.140	0.006	0.142	0.005	52.00	2.00
GP Length 100	0.141	0.006	0.147	0.007	100.00	7.75
NSGA-II Complexity	0.196	0.042	0.195	0.042	36.50	30.50
NSGA-II Visitation Length	0.160	0.024	0.158	0.024	34.00	20.00
NSGA-II Tree Size	0.154	0.048	0.157	0.051	32.50	21.25
NSGA-II Variables	0.139	0.003	0.141	0.003	86.00	28.00
Housing						
GP Length 20	0.192	0.014	0.198	0.017	24.00	1.00
GP Length 50	0.153	0.017	0.211	0.055	53.00	3.00
GP Length 100	0.132	0.022	0.202	0.090	102.00	6.75
NSGA-II Complexity	0.146	0.037	0.183	0.043	82.50	45.75
NSGA-II Visitation Length	0.157	0.064	0.198	0.033	48.50	52.25
NSGA-II Tree Size	0.152	0.060	0.192	0.036	60.50	42.00
NSGA-II Variables	0.139	0.028	0.197	0.064	102.00	8.00
Chemical						
GP Length 20	0.272	0.020	0.432	0.112	24.00	1.00
GP Length 50	0.214	0.025	0.329	0.197	54.00	2.00
GP Length 100	0.195	0.025	0.343	0.281	102.00	7.00
NSGA-II Complexity	0.209	0.025	0.270	0.094	82.00	39.75
NSGA-II Visitation Length	0.221	0.029	0.360	0.179	59.50	34.00
NSGA-II Tree Size	0.237	0.035	0.373	0.188	44.00	34.25
NSGA-II Variables	0.211	0.030	0.312	0.207	102.00	4.00
Tower						
GP Length 20	0.158	0.029	0.159	0.033	24.00	1.00
GP Length 50	0.138	0.026	0.141	0.034	53.00	3.00
GP Length 100	0.124	0.021	0.131	0.028	101.50	7.75
NSGA-II Complexity	0.127	0.017	0.128	0.022	58.00	42.75
NSGA-II Visitation Length	0.132	0.015	0.131	0.019	41.50	52.25
NSGA-II Tree Size	0.141	0.019	0.138	0.020	32.00	42.00
NSGA-II Variables	0.134	0.039	0.141	0.041	100.50	8.00

Table 7 Analysis of the functions in the best models in terms of the subtree size affected by the symbol grouped into three categories (trigonometric: sin, cos, tan—exponential: exp, log—power: x^2, $\sqrt{x}$)

	Trigonometric		Exponential		Power	
	Median	IQR	Median	IQR	Median	IQR
Breiman						
GP Length 20	0.00	2.00	2.00	6.00	0.00	2.75
GP Length 50	21.00	39.25	15.00	23.50	9.50	35.75
GP Length 100	123.00	139.50	87.50	70.50	38.00	82.75
NSGA-II Complexity	0.00	0.00	0.00	0.00	0.00	0.00
NSGA-II Visitation Length	0.00	5.50	10.00	10.00	0.00	0.00
NSGA-II Tree Size	0.00	1.50	8.00	11.00	0.00	4.75
NSGA-II Variables	151.50	182.00	96.50	130.50	79.50	113.00
Friedman						
GP Length 20	4.50	5.50	0.00	2.75	3.00	7.00
GP Length 50	44.00	41.25	8.50	20.50	9.00	30.00
GP Length 100	127.00	106.25	42.50	69.00	38.50	71.25
NSGA-II Complexity	2.00	8.25	0.00	0.00	2.50	7.75
NSGA-II Visitation Length	12.00	29.50	0.00	2.00	2.50	4.00
NSGA-II Tree Size	10.50	19.75	0.00	4.00	2.00	6.75
NSGA-II Variables	214.50	203.75	51.50	81.50	79.00	68.50
Housing						
GP Length 20	4.00	4.75	4.00	15.75	0.00	6.00
GP Length 50	19.50	20.50	31.50	53.75	26.00	33.75
GP Length 100	127.00	94.50	111.50	139.75	99.00	148.50
NSGA-II Complexity	2.00	17.50	6.00	12.75	0.00	2.00
NSGA-II Visitation Length	14.50	48.25	16.00	35.75	6.50	14.00
NSGA-II Tree Size	16.50	33.50	29.50	56.00	7.00	36.50
NSGA-II Variables	203.00	179.00	120.00	126.75	109.50	145.00
Chemical						
GP Length 20	0.00	2.00	0.00	0.00	0.00	6.00
GP Length 50	12.00	23.25	1.00	8.75	8.00	21.25
GP Length 100	58.00	91.50	24.00	65.75	57.00	68.75
NSGA-II Complexity	0.00	1.50	0.00	0.00	0.00	4.00
NSGA-II Visitation Length	0.00	22.50	0.00	0.00	5.00	51.75
NSGA-II Tree Size	0.00	1.50	0.00	0.00	12.50	36.75
NSGA-II Variables	252.50	256.25	84.00	101.75	149.50	140.00
Tower						
GP Length 20	0.00	3.50	0.00	0.00	0.00	2.00
GP Length 50	14.00	27.00	7.50	20.00	7.50	14.00
GP Length 100	61.50	112.75	42.00	78.25	38.50	92.50
NSGA-II Complexity	0.00	0.00	0.00	0.00	0.00	0.00
NSGA-II Visitation Length	6.00	19.00	2.00	12.50	0.00	4.00
NSGA-II Tree Size	6.00	27.75	3.00	24.50	0.00	7.75
NSGA-II Variables	513.00	442.25	122.50	146.00	109.50	156.75

objective genetic programming has been performed by utilizing NSGA-II with slight adaptations to make it suitable for symbolic regression. Furthermore, we defined a new complexity measure that combines syntactical information about the evolved trees and the semantics of the occurring symbols.

Among the standard genetic programming algorithms the one with the strictest size constraints worked best on the artificial problems, both in terms of the accuracy and simplicity of the models. However, this is only the case if the length constraint is large enough to generate models that could explain the data reasonably well. This picture changes when comparing the results obtained on noisy problems, where standard GP with larger size constraints works better. This indicates that the optimal length constraint is problem dependent and cannot be known a-priori, thus multiple values have to be tested during modeling.

Switching from single-objective to multi-objective genetic programming removes the necessity for specifying a length constraint, because the complexity is implicitly optimized during the algorithm execution. Additionally, we demonstrated that by including semantics of the function symbols contained in the models, the algorithm's ability to determine the necessary complexity to model the data is strengthened without worsening the accuracy of the evolved models.

Acknowledgements The work described in this paper was done within the COMET Project Heuristic Optimization in Production and Logistics (HOPL), #843532 funded by the Austrian Research Promotion Agency (FFG).

References

Affenzeller M, Winkler S, Kronberger G, Kommenda M, Burlacu B, Wagner S (2014) Gaining deeper insights in symbolic regression. In: Riolo R, Moore JH, Kotanchek M (eds) Genetic programming theory and practice XI. Genetic and evolutionary computation. Springer, New York

Breiman L, Friedman J, Stone CJ, Olshen RA (1984) Classification and regression trees. CRC Press, Boca Raton

Deb K, Pratap A, Agarwal S, Meyarivan T (2002) A fast and elitist multiobjective genetic algorithm: NSGA-II. IEEE Trans Evolut Comput 6(2):182–197

Dignum S, Poli R (2008) Operator equalisation and bloat free gp. In: Genetic programming. Springer, Berlin, pp 110–121

Friedman JH (1991) Multivariate adaptive regression splines. Ann Stat 19(1):1–67. https://projecteuclid.org/euclid.aos/1176347963

Keijzer M, Foster J (2007) Crossover bias in genetic programming. In: Genetic programming. Springer, Berlin, pp 33–44

Koza JR (1992) Genetic programming: on the programming of computers by means of natural selection. MIT Press, Cambridge, MA

Luke S (2000) Two fast tree-creation algorithms for genetic programming. IEEE Trans Evolut Comput 4(3):274–283

Luke S, Panait L (2002) Lexicographic Parsimony Pressure. In: Langdon WB, Cantú-Paz E, Mathias K, Roy R, Davis D, Poli R, Balakrishnan K, Honavar V, Rudolph G, Wegener J, Bull L, Potter MA, Schultz AC, Miller JF, Burke E, Jonoska N (eds) Proceedings of the genetic and evolutionary computation conference (GECCO'2002). Morgan Kaufmann Publishers, San Francisco, CA, pp 829–836

Poli R (2010) Covariant Tarpeian method for bloat control in genetic programming. Genet Program Theory Pract VIII 8:71–90

Poli R, Langdon WB, McPhee NF (2008) A field guide to genetic programming. Published via http://lulu.com and freely available at http://www.gp-field-guide.org.uk

Silva S, Costa E (2009) Dynamic limits for bloat control in genetic programming and a review of past and current bloat theories. Genet Program Evolvable Mach 10(2):141–179

Smits GF, Kotanchek M (2005) Pareto-front exploitation in symbolic regression. In: Genetic programming theory and practice II. Springer, Berlin, pp 283–299

Srinivas N, Deb K (1994) Multiobjective optimization using nondominated sorting in genetic algorithms. Evol Comput 2(3):221–248

Vanneschi L, Castelli M, Silva S (2010) Measuring bloat, overfitting and functional complexity in genetic programming. In: Proceedings of the 12th annual conference on genetic and evolutionary computation. ACM, New York, pp 877–884

Vladislavleva EJ, Smits GF, Den Hertog D (2009) Order of nonlinearity as a complexity measure for models generated by symbolic regression via Pareto genetic programming. IEEE Trans Evol Comput 13(2):333–349

Wagner S (2009) Heuristic optimization software systems - modeling of heuristic optimization algorithms in the heuristiclab software environment. Ph.D. thesis, Institute for Formal Models and Verification, Johannes Kepler University, Linz

White DR, McDermott J, Castelli M, Manzoni L, Goldman BW, Kronberger G, Jaskowski W, O'Reilly UM, Luke S (2013) Better GP benchmarks: community survey results and proposals. Genet Program Evol Mach 14(1):3–29. doi: 10.1007/s10710-012-9177-2

Learning Heuristics for Mining RNA Sequence-Structure Motifs

Achiya Elyasaf, Pavel Vaks, Nimrod Milo, Moshe Sipper, and Michal Ziv-Ukelson

Abstract The computational identification of conserved motifs in RNA molecules is a major—yet largely unsolved—problem. Structural conservation serves as strong evidence for important RNA functionality. Thus, comparative structure analysis is the gold standard for the discovery and interpretation of functional RNAs.

In this paper we focus on one of the functional RNA motif types, sequence-structure motifs in RNA molecules, which marks the molecule as targets to be recognized by other molecules.

We present a new approach for the detection of RNA structure (including pseudo-knots), which is conserved among a set of unaligned RNA sequences. Our method extends previous approaches for this problem, which were based on first identifying conserved stems and then assembling them into complex structural motifs. The novelty of our approach is in simultaneously preforming both the identification and the assembly of these stems. We believe this novel unified approach offers a more informative model for deciphering the evolution of functional RNAs, where the sets of stems comprising a conserved motif co-evolve as a correlated functional unit.

Since the task of mining RNA sequence-structure motifs can be addressed by solving the maximum weighted clique problem in an n-partite graph, we translate the maximum weighted clique problem into a state graph. Then, we gather and define domain knowledge and low-level heuristics for this domain. Finally, we learn hyper-heuristics for this domain, which can be used with heuristic search algorithms (e.g., A*, IDA*) for the mining task.

The hyper-heuristics are evolved using HH-Evolver, a tool for domain-specific, hyper-heuristic evolution. Our approach is designed to overcome the computational limitations of current algorithms, and to remove the necessity of previous assumptions that were used for sparsifying the graph.

This is still work in progress and as yet we have no results to report. However, given the interest in the methodology and its previous success in other domains we are hopeful that these shall be forthcoming soon.

A. Elyasaf (✉) • P. Vaks • N. Milo • M. Sipper • M. Ziv-Ukelson
Department of Computer Science, Ben-Gurion University, Beer-Sheva 84105, Israel
e-mail: achiya.e@gmail.com; pavel.vaks@gmail.com; milo.nimrod@gmail.com;
sipper@cs.bgu.ac.il; michaluz@cs.bgu.ac.il

© Springer International Publishing Switzerland 2016
R. Riolo et al. (eds.), *Genetic Programming Theory and Practice XIII*,
Genetic and Evolutionary Computation, DOI 10.1007/978-3-319-34223-8_2

21

Keywords Genetic algorithms • Genetic programming • Hyper heuristic

1 Introduction

1.1 RNA Structural Motif Discovery

RNA is a biological macromolecule which, like DNA, is constructed of four letter alphabet (A, C, G and U). RNA has many roles in biological mechanisms, some of which we describe below.

Over the last few years non-coding RNAs (ncRNAs) have been recognized as a highly abundant class of RNAs that do not code for proteins but nevertheless are functional in many biological processes, including localization, replication, translation, degradation, regulation, and stabilization of biological macromolecules (Mandal and Breaker 2004).

1.2 Biological Preliminaries and Definitions

An RNA molecule is defined by a *sequence* of letters (called bases) and a set of pairings between its bases. The base *C* typically pairs with *G*, *A* typically pairs with *U*, and another weaker pairing can occur between *G* and *U*. This base-paired structure is called the *secondary structure* of the RNA. Paired bases almost always occur in a nested fashion. Informally, this means that if we draw arcs over an RNA sequence connecting base pairs, none of the arcs cross each other. When non-nested base pairs occur, they are called *pseudoknots* (see Fig. 1). Most of current RNA sequence-structure analysis algorithms ignore pseudoknots. This is done mostly in order to simplify the problem, due to the fact that prediction of structure while allowing pseudoknots is NP hard (Akutsu 2000). In nature, there are important examples of RNA sequence-structure motifs that include pseudoknots (Staple and Butcher 2005; Brierley et al. 2008).

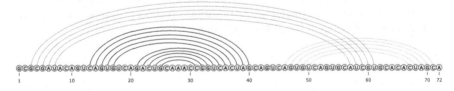

Fig. 1 An RNA sequence and its structure (defined by the *arcs*). In the figure there are three stems (*marked red, green, and blue*). The *green and red stems* cross each other, indicating that the structure of the exemplified RNA contains a pseudoknot (Color figure online)

In RNA studies, structural conservation serves as a strong evidence for essential RNA functionality. Thus, comparative structure analysis is the gold standard for the discovery and interpretation of functional RNAs. In particular, an important and well-studied problem is that of RNA motif discovery. In molecular biology, a motif is a sequence pattern that is widespread and has, or is conjectured to have, a biological significance. In the case of functional RNAs, the sought motifs are patterns combining sequence and structural features, thus denoted RNA sequence-structure motifs.

In RNA motif discovery, a natural building block is a stem (a local non-crossing set of base pairs). Therefore, some approaches for the mining of RNA motifs are based on first identifying sets of stems that are widespread, and then combining the stems to form more complex structural motif patterns. In this paper we present a novel approach that addresses both problems *simultaneously* using learning of heuristic functions and search techniques. Given a set of n RNA molecules we seek a set of similar stems, each shared by a minimum of $k \leq n$ molecules. Using A^* algorithm along with a learned heuristic function, we mine the n molecules both in sequence and complex structural characteristics. The measure we use for scoring the similarity between two stems models the typical changes caused by evolution.

1.3 Heuristic Search

Heuristic search algorithms are strongly based on the notion of approximating the cost of the cheapest path from a given configuration (or *state*) to the problem's solution (or *goal*). Such approximations are found by means of a computationally efficient function, known as a *heuristic function*. By applying such a function to states reachable from the current one considered, it becomes possible to select more-promising alternatives earlier in the search process, possibly reducing the amount of search effort (typically measured in number of states expanded) required to solve a given problem. The putative reduction is strongly tied to the quality of the heuristic function used: employing a perfect function means simply "strolling" onto the solution (i.e., no search de facto), while using a bad function could render the search less efficient than totally uninformed search, such as breadth-first search (BFS) or depth-first search (DFS).

A heuristic function is said to be *admissible* if it never overestimates the cost to the goal. Thus, the higher the heuristic value, the closer it is to the true cost of the cheapest path to goal. Using an admissible heuristic with an optimal search algorithm (e.g. A^* or iterative deepening A^*, IDA*, Hart et al. 1968; Korf 1985) guarantees that any solution found will be optimal, i.e., with minimal solution length.

Combining several heuristics to get a more accurate one is considered one of the most difficult problems in contemporary heuristics research (Samadi et al. 2008; Burke et al. 2010). Of course, if all of the heuristic functions are admissible and an optimal solution is what we are looking for, then we could use the max heuristic (which takes the heuristic function with the maximal value). However, when we do not need to guarantee optimality or when we do not use only admissible heuristics, a good solution can be found in a more efficient way.

1.4 Hyper Heuristics

Within combinatorial optimization, the term hyper-heuristics was first used in 2000 (Cowling et al. 2000) to describe heuristics to choose heuristics. This definition of hyper-heuristics was expanded later (Burke et al. 2010) to refer to an automated methodology for selecting or generating heuristics to solve hard computational search problems. In the process of hyper-heuristics learning, heuristics and domain knowledge are used as building blocks. These heuristics can be of high level, usually complex and memory-consuming (e.g., landmarks and pattern databases), or low-level domain knowledge and heuristics that are usually intuitive and straightforward to implement and compute.

Hyper-heuristics have been applied in many research fields, among them:

- Classical planning (Yoon et al. 2008; Fawcett et al. 2011; Hoffmann and Nebel 2001).
- Classical NP-Complete domains, e.g., personnel scheduling (Burke et al. 2003), traveling salesman (Oltean 2005), and vehicle routing (Garrido and Rojas 2010).
- Classical AI domains and puzzles, e.g., the Rush Hour puzzle (Hauptman et al. 2009), the game of FreeCell (Elyasaf et al. 2012), and the Tile Puzzle (Arfaee et al. 2010; Samadi et al. 2008).

The growing research interest in techniques for automating the design of heuristic search methods motivates the search for automatic systems for generating hyper-heuristics.

1.5 Our Approach: Learning Hyper Heuristics for the Task of Mining RNA Sequence-Structure Motifs

In this paper, we present a novel way for mining RNA sequence-structure motifs. We translate the problem into a search graph and devise 97 heuristics for this domain. Next, we use these heuristics as building blocks for learning hyper heuristics using HH-Evolver—an evolutionary algorithm system designed for this type of learning.

This is still work in progress and as yet we have no results to report. However, given the interest in the methodology and its previous success in other domains we are hopeful that these shall be forthcoming soon.

The contributions of this work are as follows:

1. We present a novel approach the task of mining RNA sequence-structure motifs with possible pseudoknots.
2. While previous approaches artificially divide the mining task into conserved stem identification followed by the assembly of such stems into complex structural motifs, we present the *first* approach for preforming the two tasks simultaneously. We believe this unified approach offers a more informative model for deciphering the evolution of functional RNAs, where the sets of stems comprising a conserved motif co-evolve as a correlated functional unit.
3. By using a more efficient search we are able to use a denser graph with more nodes (stems) and edges (relations between stems). Thus, we are able to remove the necessity of previous algorithms to sparsify the graph, at the expense of sensitivity, by imposing assumptions and rigid limitations.
4. The use of hyper heuristics enables the mining algorithm to find the exact conditions (i.e., biological indicators) regarding *when* to apply each heuristic, or combinations.
5. We push the limit of what has been done with evolution further mining RNA sequence-structure motifs one of the most difficult domains to which evolutionary algorithms have been applied to date.
6. Along the way we devise several novel heuristics for this domain. The methodology of creating these heuristics could be applied to other biological domains and other.
7. By devising novel heuristics and evolving them into hyper-heuristics, we continue our previous presentation of a new framework for solving many (non-admissible) heuristic problems, which proved to be efficient and successful.
8. We transform a non-search domain into a state graph, and thus strengthen the bridge between the learning and search community to the biological world.

The rest of the paper is organized as follows: In the next section we examine previous and related work. In Sect. 3 we describe our method. Finally, we end with concluding remarks and future work in Sect. 4.

2 Previous Work

2.1 Mining Common Structure Among a Set of Unaligned RNA Sequences

Several approaches exist for identifying common structure among a given set of RNA molecules. The first approach (Pederson et al. 2006; Hofacker et al. 2002; Washietl and Hofacker 2004) is to align sequences using standard multiple sequence alignment tools (e.g., ClustalW, Thompson et al. 1994), then use structure-conserving mutations for the inference of a consensus structure. However, in order for this approach to work the multiple sequence-alignment step needs to rely on a very well-conserved sequence, which is rarely the case with swiftly evolving ncRNAs.

The second approach applies Sankoff's "Simultaneous Alignment with Folding (SAF)"—an algorithm designed to simultaneously align a set of co-functional RNA molecules and predict their common secondary structure. However, the algorithm requires extensive computational resources both in time $(O(n^6))$ and memory $(O(n^4))$ (Backofen et al. 2011). Thus, current implementations (Mathews and Turner 2002; Havgaard et al. 2005; Will et al. 2007; Siebert and Backofen 2005; Hofacker et al. 2004; Torarinsson et al. 2007) are either restricted in the input size or apply a restricted application of the SAF approach.

The third approach, denoted as the "pre-folding" approach, is applied when no helpful level of sequence conservation is observed. It excludes the sequence alignment step, predicts secondary structures for each sequence (or a sub-group of sequences) separately, and directly aligns the structures. Because of the nested branching nature of RNA structures, these are represented as trees. Tree comparison and tree alignment models in the context of detecting conserved RNA structure have been proposed and implemented in Hofacker et al. (1994), Sczyrba et al. (2003), Hochsmann et al. (2003), Wang and Zhang (2001), and Milo et al. (2013). Predicting a global secondary structure from a single molecule is still error, where the best approaches may yield up to 75 % accuracy (Do et al. 2006).

Due to computational limitations the above methods are generally restricted to finding conserved sequence-structure motifs without considering pseudoknots. A leading approach that removes this restriction is the one proposed by Ji et al. (2004). Here, rather then trying to predict a *global* structure, i.e., one that is common to the (full-length) input sequences, *local* common structure is sought, i.e., a set of substrings that share a similar predicted structure, where each substring belongs to a different input sequence.

Ji et al. apply a multi-stage approach to identify common structure among a given set of n sequences. During a preprocessing stage, an n-partite undirected weighted stem-graph is constructed, by first extracting local stems from each of the sequences. The extracted stems serve as the nodes of the graph, and are partitioned by their sequence of origin. Weighted edges are constructed between pairs of nodes representing the similarity between stems that were extracted from different sequences. Stem similarity is computed by using a single pre-defined scoring function. Ji et al. sparsify this stem-graph by using fixed thresholds on the scoring function and the stability of the stems. In addition, they require the existence of conserved sequential regions of a fixed size, termed *anchors*.

In the next stage, conserved stems are identified. This is done by mining all cliques in the stem-graph spanning a minimum of $k \leq n$ sequences.

This is followed by a final stage, where stem-cliques are combined to form complex motifs. This is achieved by assembling as many compatible cliques as possible according to topological order and evaluating the plausibility of a proposed structure by the significance of its members.

In this paper we extend the work of Ji et al. By applying evolution-based learning, we remove the necessity of the aforementioned single scoring function and the stringent cutoff thresholds. Our search is more informed, and computes intra-clique optimizations within the wide context of topological inter-clique assembly considerations. This allows us to speed up the search without the loss of sensitivity. In fact, sensitivity can now be increased, as the more efficient search allows us to get rid of the preprocessing sparsification used by Ji et al. We use evolution to combine multiple scoring functions and dynamic thresholds.

2.2 Learning Hyper Heuristics

2.2.1 Learning Hyper Heuristics for Planning Systems

Planning systems are strongly based on the notion of heuristics (e.g., Bonet and Geffner 2005; Hoffmann and Nebel 2001). However, relatively little work has been done on *evolving* heuristics for planning.

Aler et al. (2002) (see also Aler et al. 1998, 2001) proposed a multi-strategy approach for learning heuristics, embodied as ordered sets of control rules (called *policies*), for search problems in AI planning. Policies were evolved using a GP (Genetic Programming) based system called EvoCK (Aler et al. 2001), whose initial population was generated by a specialized learning algorithm, called Hamlet (Borrajo and Veloso 1997). Their hybrid system, Hamlet-EvoCK, outperformed each of its sub-systems on two benchmark problems often used in planning: Blocks World and Logistics (solving 85 and 87 % of the problems in these domains respectively). Note that both these domains are considered relatively easy (e.g., compared to Rush Hour and FreeCell, mentioned above), as evidenced by the fact that the last time they were included in an IPC (International Planning Competition) was in 2002.

Levine and Humphreys (2003), and later Levine et al. (2009), also evolved policies and used them as heuristic measures to guide search for the Blocks World and Logistic domains. Their system, L2Plan, included rule-level genetic programming (for dealing with entire rules), as well as simple local search to augment GP crossover and mutation. They demonstrated some measure of success in these two domains, although hand-coded policies sometimes outperformed the evolved ones.

2.2.2 Learning Hyper Heuristics for Specific Domains

Samadi et al. (2008) used artificial neural networks (ANNs) (Mitchell 1999) for learning combinations of heuristics for the sliding-tile puzzle and the 4-peg Towers of Hanoi. They used pattern databases (PDBs) (Korf 1997) and weighted PDBs as input signals for the ANN.

Arfaee et al. (2010) also used ANNs for learning hyper heuristics for several domains, however, in addition to the use of small PDBs as input signals, Arfaee et al. used low-level heuristics and domain knowledge as well. For the sliding-tile puzzle, for example, they used the following additional signals: the number of out-of-place tiles, the position of the blank, the number of tiles not in the correct row, and the number of tiles not in the correct column.

Hauptman et al. (2009) and later Elyasaf et al. (2012) evolved heuristics for the Rush Hour puzzle and the game of FreeCell (respectively). They compared the performance of different types of hyper heuristics on these domains. In both cases, evolved hyper heuristics (along with heuristic search) greatly reduced the number of nodes required to solve instances of the domains, compared to standard methods and previous solvers. In this paper we use their method for learning hyper heuristics for the problem of mining RNA sequence-structure motifs.

3   Method

As explained in Sect. 2.1, we extend the work of Ji et al. (2004). The overall scheme of our method is summarized by the following major steps:

- Cast the problem of mining RNA sequence-structure motifs as one of maximum weighted clique in an n-partite graph.
- Translate the maximum weighted clique problem into a search graph, where each state consists of the set of cliques identified so far.
- Gather and define domain knowledge and low-level heuristics for this domain.
- Learn hyper-heuristics for this domain, which can be used with heuristic search algorithms (e.g., A*, IDA*) for the mining task.

In their paper, Ji et al. (2004) list several heuristics and techniques that already exist for the task of mining RNA sequence-structure motifs and for the task of finding maximum weighted cliques, however they note that they are *"not aware of any effective optimization or heuristic algorithm feasible for our problem"*. For this reason they used a fixed scoring function and harsh pruning thresholds (see Sect. 2.1).

Since there are many heuristics, both for the task of mining RNA sequence-structure motifs and for the task of finding maximum weighted cliques, we believe that learning algorithms can find a combination of these heuristics that outperforms the handcrafted scoring formula used by Ji et al.

The key differences between our approach and Ji et al. are: (1) We use learning algorithms with *many* heuristics along with domain knowledge to avoid the use of a single formula; (2) we let evolution guide the search and remove the need to rely on fixed thresholds and anchoring; (3) we extend the stem extraction module by using the approach of Milo et al. (2014), which allows us find a wider range of stem types; (4) we perform both mining tasks (identification of conserved stems and the assembly of conserved stems into complex motif structures) simultaneously.

3.1 Casting the Problem of Mining RNA Sequence-Structure Motifs as One of Maximum Weighted Clique in an n-Partite Graph

Given a set of n molecules, we extend the stem extraction module of Ji et al. by using the extraction method described in Milo et al. (2014), allowing us to extract more stems from the molecules. Next, we construct an n-partite undirected weighted graph to represent the stems and their relations. Each vertex represents a stem, and the graph is partitioned into n parts. Each part comprises the stems from one molecule. Only stems (or vertices) from different parts can be connected. We call the edges between the stems *stem edges*, to differentiate between these edges and the edges described in Sect. 3.2. While Ji et al. sparsify this graph using rigid thresholds, we do not sparsify the graph at this stage and connect *all* stems of different partitions.

For each stem edge we define several features (dubbed stem edge features, or SEF), described in Sect. 3.3.1. For each feature it is possible to define a hard threshold set by the user, or let the learning algorithm set the threshold value.

3.2 Converting the Maximum Weighted Clique Problem into a State Graph

Once the n-partite graph is built, Ji et al. (2004) extract all possible stem-cliques, a task for which the worst-case time complexity is $O(m^n)$, where m is the maximum number of stems examined in one sequence and n is the number of total RNA sequences. The average run-time is much less than the worst case, due to the thresholds and pruning. However, the time could still be impractical in many cases, and furthermore the sparsification comes at the expense of data-mining sensitivity. Moreover, if the goal is to mine the input and find significantly similar structures among the molecules, one does not need to exhaustively consider all possible cliques. If we are able to find the most important cliques first, we can stop the search at an earlier phase. This is the reason that we now turn to heuristic search algorithms such as A*.

In order to search for cliques in an A* manner, we first define a new search graph, where each state contains the set of cliques (and partial cliques) found so far. The edges represent either the start of a new clique or adding a stem to an existing clique. As opposed to standard search domains, here we do not know which state is the goal state (i.e., the state with the most important cliques). We will discuss this point further in Sect. 3.4.3.

We now turn to describe the domain knowledge and low-level heuristics gathered for the search algorithm.

3.3 Gather and Define Domain Knowledge and Low-Level Heuristics for this Domain

There are three different types of heuristics we use:

1. *SEF heuristics.* heuristics that are derived from the stem edge features. Recall that each search state represents the cliques founds so far and that each clique contains several stem edges. With this notion, we define seven heuristics for each stem edge feature, f:

   (a) Return the minimal/maximal/average/median f value of all stem edges of the state.
   (b) Compute average f value per clique; return average of all average values.
   (c) Compute median f value per clique; return average of all median values.
   (d) Compute average f value per clique; return median of all average values.

   The full SEF list is described below.

2. *Topological relation heuristics.* heuristics that measure the topology preservation between stems from different cliques using the *topological relations*. We have two heuristics of this type: The first is described in Milo et al. (2014), and the second is a heuristic version of the structure assembly algorithm described in Ji et al. (2004).

3. *Clique-specific heuristics.* standard clique heuristics such as maximal, minimal, average, median node degree.

All of our heuristics preserve the basic rule of heuristic functions: the lower the value, the closer we are to the goal. Towards this end we change the heuristic value h to $1/h$, where applicable.

There is a fourth type of heuristics—classic search heuristics (e.g., pattern database)—however, previous work has shown that using domain knowledge-based heuristics provides good solutions.

The distribution of the heuristic types is described in Table 1.

Table 1 Distribution of heuristic types

Heuristic type	Number of heuristics
SEF	91 (7 heuristics * 13 features = 91)
Topological relation heuristics	2
1 Clique-specific heuristics	4
Total	97

3.3.1 Stem Edge Features

As described in Sect. 2.1, Ji et al. (2004) defined a fixed equation for describing similarity between two stems, which was a combination of five features. We use these features, as well as the equation, as part of our stem edge features (SEF), along with additional features described below.

Some of our features are added twice: once as is (i.e., $f_l(i_x, j_y)$), and once divided by the energy, using the formula $\bar{f}_l(i_x, j_y) = \frac{2 \cdot f_l(i_x, j_y)}{2 + r_x(i) + r_y(j)}$ (as described by Ji et al.).

The complete list of our features is described in Table 2.

Table 2 The list of features

Feature	Origin	Description
f_1 : Helix length	[1]	Number of base-pairs in the stem
$f_2 : f_2 = \bar{f}_1$	[1]	[*]
f_3 : Helix sequence	[1]	The sequence of bases in the stem
$f_4 : f_4 = \bar{f}_3$	[1]	[*]
f_5 : Loop sequence	[1]	The sequence of letters between the innermost base-pair in the stem
$f_6 : f_6 = \bar{f}_5$	[1]	[*]
f_7 : Stem stability	[1]	The free energy value of the stem
$f_8 : f_8 = \bar{f}_7$	[1]	[*]
f_9 : Relative positions	[1]	The position of the left base in the outermost base-pair in the stem (relative to the sequence)
$f_{10} : f_{10} = \bar{f}_9$	[1]	[*]
f_{11} : Ji. et al. Similarity	[1]	$\frac{2 \cdot \sum_{l=1,3,5,7} \{w_l \cdot f_l(i_x, j_y)\}}{2 + r_x(i) + r_y(j)}$
f_{12} : Context	[3]	The shift between both helix counterparts of a stem in the anchor or non-anchor region
f_{13} : StemSearch similarity	[2]	Similarity score used by StemSearch, determined by structural and sequential similarity

[1]—Features taken from Ji et al. (2004)
[2]—Features taken from Milo et al. (2014)
[3]—Features designed by us
[*]—$\bar{f}_l(i_x, j_y) = \frac{2 \cdot f_l(i_x, j_y)}{2 + r_x(i) + r_y(j)}$

3.4   Learning Hyper Heuristics Using HH-Evolver

Combining several heuristics to get a more accurate one is considered one of the most difficult problems in contemporary heuristics research (Samadi et al. 2008; Burke et al. 2010).

This task typically involves solving three major sub-problems:

1. How to combine heuristics by *arithmetic* means, e.g., by summing their values or taking the maximal value.
2. Finding exact conditions (i.e., *logic* functions) regarding *when* to apply each heuristic, or combinations thereof—some heuristics may be more suitable than others when dealing with specific state configurations.
3. Finding the proper set of domain configurations in order to facilitate the learning process while avoiding pitfalls such as overfitting.

The problem of combining heuristics is difficult mainly because it entails traversing an extremely large search space of possible numeric combinations, logic conditions, and state configurations. To tackle this problem we turn to *evolution*.

As previously mentioned, we use the learning method of Hauptman et al. (2009) and Elyasaf et al. (2012) for the mining problem. For this task we use their tool, *HH-Evolver* (Elyasaf and Sipper 2013), a hyper-heuristic generator for search domains. The HH-Evolver system receives as input: a domain, several heuristics for the domain, and a dataset of domain instances to be used partly as training set and partly as test set. HH-Evolver generates a population of random hyper-heuristics and trains them over generations against the training set. When used with a heuristic search algorithm, the individuals are required to produce near-optimal solutions to the instances encountered. The search-size (i.e., the number of states encountered during the search) should be small as well.

In order to properly solve the aforementioned sub-problems, we use three different HH-Evolver genomic representations—Standard GA, GP, and policies— all of which are thoroughly described below. These representations use the heuristics given as input to HH-Evolver.

3.4.1   The Hyper Heuristic-Based Genome

We use three different genomic representations. All of these representation were used by Hauptman et al. (2009) and Elyasaf et al. (2012).

All of our representations comprise real-value thresholds for all minimum heuristics (e.g., minimal node degree heuristic). During the search we prune nodes when one of the heuristic values exceeds its threshold.

Standard GA (Genetic Algorithm)

This type of hyper-heuristic only addresses the first problem of how to combine heuristics by arithmetic means. Each individual comprises 97 real values in the range $[0, 1]$, representing a linear combination of all 97 heuristics described above (Table 1). Specifically, the heuristic value, H, designated by an evolving individual is defined as $H = \sum_{i=1}^{97} w_i h_i$, where w_i is the ith weight specified by the genome, and h_i is the ith heuristic shown in Table 1. As results in other domains showed (Hauptman et al. 2009; Elyasaf et al. 2012), the GA proved quite successful and was therefore retained as a yardstick to measure against when we embarked upon our GP path.

GP (Genetic Programming)

As we want to embody both combinations of estimates and conditions for applying each combination, we evolve GP-trees as described in Koza (1994). The function set included the functions $\{IF, AND, OR, \leq, \geq, *, +\}$, and the terminal set included all heuristics in Table 1, as well as random numbers within the range $[0, 1]$.

Policies

The last genome used also combines estimates and application conditions, using ordered sets of control rules, or *policies*. As stated above, policies have been evolved successfully with GP to solve search problems—albeit simpler ones (e.g., Hauptman et al. 2009; Elyasaf et al. 2012).

The policies have the following structure:

$RULE_1$: IF *Condition₁* THEN *Value₁*

.

.

.

$RULE_N$: IF *Condition_N* THEN *Value_N*
$DEFAULT$: *Value_{N+1}*

where *Condition_i* and *Value_i* represent conditions and estimates, respectively.

Policies are used by the search algorithm in the following manner: The rules are ordered such that we apply the first rule that "fires" (meaning its condition is true for the current state being evaluated), returning its *Value* part. If no rule fires, the value is taken from the last (default) rule: *Value_{N+1}*. Thus individuals, while in the form of policies, are still heuristics—the value returned by the activated rule is an arithmetic combination of heuristic values, and is thus a heuristic value itself. This accords with our requirements: rule ordering and conditions control when we apply a heuristic combination, and values provide the combinations themselves.

Thus, with N being the number of rules used, each individual in the evolving population contains N *Condition* GP trees and $N + 1$ *Value* sets of weights used for computing linear combinations of heuristic values. Experimenting with several sizes of policies in different domains showed that $N = 5$ provides enough rules per individual, while avoiding cumbersome individuals with too many rules. The depth limit used for the *Condition* trees is set to 5 for the same reason.

For *Condition* GP trees, the function set included the functions $\{AND, OR, \leq, \geq\}$, and the terminal set included all heuristics in Table 1. The sets of weights appearing in *Value*s all lie within the range $[0, 1]$, and correspond to the heuristics listed in Table 1.

The genetic operators for the individuals are described thoroughly in Elyasaf et al. (2012).

3.4.2 Training and Test Sets

We use the data from the *RFAM* (Griffiths-Jones et al. 2005) database to create training and testing data sets. RFAM is a curated database of RNA molecules grouped together by common functionality (termed *RNA families*). This allows us to generate a large and diverse set of syntactic inputs for our hyper-heuristic population. This set is divided to training set and test sets, using a standard cross-validation method.

The set is generated as follow: For each family we randomly divide the members into different (overlapping) groups of different sizes. We then add to each group a random amount of noise in the range of $[0, \frac{n}{5}]$, where the noise is defined as members of a different functional family. Each generated group is used as an instance in the set.

The above technique allows us to collect different statistical metrics needed for computing the correctness of solutions (described below).

3.4.3 Fitness

The individual's fitness score preserves the rule that the higher the value, the better the individual (as opposed to heuristic functions). The score is obtained by running A* on m instances taken from the training set, with the individual used as the heuristic function. The g value is defined as the length of the path from the initial state to the current one. During the search, each individual keeps the best state found so far (i.e., the one with the lowest heuristic value). Once the search space is exhausted or the time limit (described below) is reached, the individual returns that state, designating it as the goal state.

The fitness score is then calculated, incorporating three elements:

1. `Answer's correctness (80%)`: The most important element is the correctness, hence its weight is 80% of the score. The correctness calculation is described below.
2. `Time to solution (10%)`: The time to solution is defined as the average time it took the individual to find the solution for all instances. We wish to reduce this time, and thus the element's weight is 10%. The score of the element is $1/time\_to\_solution$.
3. `Number of cliques (10%)`: Since the cliques represent the similarities between the molecules, the more correct cliques the individual finds, the better the solution is. The element's weight is 10%.

Each element is normalized to a value in the range $[0, 1]$. Towards this end we divide the element value by its maximal value, where applicable.

Correctness is calculated by the *Positive Predictive Values* (PPV) formula: $correctness = \frac{TP}{TP+FP}$, where TP is the number of times the individual marked a stem correctly in a clique (for all training instances), and FP is the number of times the individual marked a stem incorrectly in a clique (also for all training instances). In case $TP = FP = 0$, the correctness value is 0. For example, if we have three molecules from the same family and one "noisy" molecule from a different family, we increase TP by one for each stem in a clique that belongs to a molecule of the family. Consequently, we increase FP by one for each stem in a clique that belongs to the noisy molecule.

The number of instances the individuals are trained on, m, is limited by the computational resources. Yet, the user should set it to be as high as possible in order to reach convergence. Avoiding over-fitting is achieved by using a standard cross-validation method.

The minimum clique size, k, is set by the user according to the molecular families and should be in the range $[\frac{n}{2}, n]$.

3.4.4 Search Time

The time limit for a hyper-heuristic to return an answer is set initially by the user. However, during evolution, we collect the median time for finding the best state. In such a way we can reduce the time limit to be the *average time + 1 min*.

4 Concluding Remarks

We presented a new approach for the detection of RNA structure (including pseudoknots), which is conserved among a set of unaligned RNA sequences. Our method extends previous approaches that were based on first identifying conserved stems and then assembling them into complex structural motifs. The novelty of our approach is in simultaneously preforming both the identification and the assembly of these stems. We believe this novel unified approach offers a more informative model for deciphering the evolution of functional RNAs, where the sets of stems comprising a conserved motif co-evolve as a correlated functional unit.

We began by casting the problem of mining RNA sequence-structure motifs as one of maximum weighted clique in an n-partite graph of stems. Next we addressed the maximum weighted clique problem as a heuristic search problem, by translating it into a search graph, where each state consists of the set of cliques identified so far. We then defined domain knowledge and low-level heuristics for this domain. Finally, we learned hyper-heuristics for this domain, which could be used with heuristic search algorithms (e.g., A*, IDA*) for the mining task.

This paper extends a leading approach by Ji et al. (2004) with several important modifications. By applying evolution-based learning we remove the necessity of their single scoring function and the stringent cutoff thresholds. Our search is more informed and computes intra-clique optimizations within the wide context of topological inter-clique assembly considerations. This allows us to speed up the search without the loss of sensitivity. In fact, sensitivity can now be increased, as the more efficient search obviates the preprocessing sparsification used by Ji et al. We used evolution to combine multiple scoring functions and dynamic thresholds.

This is still work in progress and as yet we have no results to report. However, given the interest in the methodology and its previous success in other domains we are hopeful that these shall be forthcoming soon (servers are churning as you read these lines).

Our work may lead to several possible research directions:

- We believe our unified approach offers a more informative model for deciphering the evolution of functional RNAs, where the sets of stems comprising a conserved motif co-evolve as a correlated functional unit. We hope that our results will shed light on functional RNAs and that new biological discoveries will follow.
- The methodology presented here could be applied to other biological and non-biological domains.
- There are many heuristics and similarity measurements that are not described in this can be incorporated into our system.

Acknowledgements This research was supported by the Israel Science Foundation (grant no. 123/11 and grant no. 179/14).

References

Akutsu T (2000) Dp algorithms for rna secondary structure prediction with pseudoknots. Discrete Appl Math 104(1–3):45–62

Aler R, Borrajo D, Isasi P (1998) Genetic programming of control knowledge for planning. In: Proceedings of AIPS-98

Aler R, Borrajo D, Isasi P (2001) Learning to solve planning problems efficiently by means of genetic programming. Evol Comput 9(4):387–420

Aler R, Borrajo D, Isasi P (2002) Using genetic programming to learn and improve knowledge. Artif Intell 141(1–2):29–56

Arfaee SJ, Zilles S, Holte RC (2010) Bootstrap learning of heuristic functions. In: Proceedings of the 3rd international symposium on combinatorial search (SoCS2010), pp 52–59

Backofen R, Tsur D, Zakov S, Ziv-Ukelson M (2011) Sparse folding: time and space efficient algorithms. J Discrete Algorithms 9(1):12–31

Bonet B, Geffner H (2005) mGPT: A probabilistic planner based on heuristic search. J Artif Intell Res 24:933–944

Borrajo D, Veloso MM (1997) Lazy incremental learning of control knowledge for efficiently obtaining quality plans. Artif Intell Rev 11(1–5):371–405

Brierley I, Gilbert RC, Pennell S (2008) Pseudoknots and the regulation of protein synthesis. Biochem Soc Trans 36(4):684–689

Burke EK, Kendall G, Soubeiga E (2003) A tabu-search hyperheuristic for timetabling and rostering. J Heuristics 9(6):451–470. http://dx.doi.org/10.1023/B:HEUR.0000012446.94732.b6

Burke EK, Hyde M, Kendall G, Ochoa G, Ozcan E, Woodward JR (2010) A classification of hyperheuristic approaches. In: Gendreau M, Potvin J (eds) Handbook of meta-heuristics, 2nd edn. Springer, Berlin, pp 449–468

Cowling PI, Kendall G, Soubeiga E (2000) A hyperheuristic approach to scheduling a sales summit. In: Burke EK, Erben W (eds) PATAT. Lecture notes in computer science, vol 2079. Springer, Berlin, pp 176–190. doi:10.1007/3-540-44629-X_11

Do CB, Woods DA, Batzoglou S (2006) Contrafold: RNA secondary structure prediction without physics-based models. Bioinformatics 22(14):e90–e98

Elyasaf A, Sipper M (2013) Hh-evolver: a system for domain-specific, hyper-heuristic evolution. In: Proceedings of the 15th annual conference companion on genetic and evolutionary computation GECCO '13 companion. ACM, New York, pp 1285–1292. doi:10.1145/2464576.2482707. http://doi.acm.org/10.1145/2464576.2482707

Elyasaf A, Hauptman A, Sipper M (2012) Evolutionary design of FreeCell solvers. IEEE Trans Comput Intell AI Games 4(4):270–281. doi:10.1109/TCIAIG.2012.2210423. http://ieeexplore.ieee.org/xpls/abs_all.jsp?arnumber=6249736

Fawcett C, Karpas E, Helmert M, Roger G, Hoos H (2011) Fd-autotune: domain-specific configuration using fast-downward. In: Proceedings of ICAPS-PAL 2011

Garrido P, Rojas MCR (2010) DVRP: a hard dynamic combinatorial optimisation problem tackled by an evolutionary hyper-heuristic. J Heuristics 16(6):795–834. http://dx.doi.org/10.1007/s10732-010-9126-2

Griffiths-Jones S, Moxon S, Marshall M, Khanna A, Eddy SR, Bateman A (2005) RFAM: annotating non-coding RNAS in complete genomes. Nucleic Acids Res 33(suppl 1):D121–D124

Hart PE, Nilsson NJ, Raphael B (1968) A formal basis for heuristic determination of minimum path cost. IEEE Trans Syst Sci Cybern 4(2):100–107

Hauptman A, Elyasaf A, Sipper M, Karmon A (2009) GP-Rush: using genetic programming to evolve solvers for the Rush Hour puzzle. In: GECCO'09: Proceedings of 11th annual conference on genetic and evolutionary computation conference. ACM, New York, pp 955–962. doi:10.1145/1569901.1570032. http://dl.acm.org/citation.cfm?id=1570032

Havgaard J, Lyngso R, Stormo G, Gorodkin J (2005) Pairwise local structural alignment of RNA sequences with sequence similarity less than 40%. Bioinformatics 21(9):1815–1824

Hochsmann M, Toller T, Giegerich R, Kurtz S (2003) Local similarity in RNA secondary structures. In: Proceedings of the IEEE computer society conference on bioinformatics, Citeseer, p 159

Hofacker I, Fontana W, Stadler P, Bonhoeffer L, Tacker M, Schuster P (1994) Fast folding and comparison of RNA secondary structures. Monatshefte fur Chemie/Chemical Monthly 125(2):167–188

Hofacker I, Fekete M, Stadler P (2002) Secondary structure prediction for aligned RNA sequences. J Mol Biol 319:1059–1066

Hofacker I, Bernhart S, Stadler P (2004) Alignment of RNA base pairing probability matrices. Bioinformatics 20(14):2222–2227

Hoffmann J, Nebel B (2001) The FF planning system: fast plan generation through heuristic search. J Artif Int Res 14(1):253–302. http://dl.acm.org/citation.cfm?id=1622394.1622404

Ji Y, Xu X, Stormo GD (2004) A graph theoretical approach for predicting common rna secondary structure motifs including pseudoknots in unaligned sequences. Bioinformatics 20(10):1591–1602

Korf RE (1985) Depth-first iterative-deepening: an optimal admissible tree search. Artif Intell 27(1):97–109

Korf RE (1997) Finding optimal solutions to Rubik's cube using pattern databases. In: Proceedings of the fourteenth national conference on artificial intelligence and ninth conference on innovative applications of artificial intelligence, AAAI'97/IAAI'97, AAAI Press, pp 700–705

Koza JR (1994) Genetic programming II: automatic discovery of reusable programs. MIT Press, Cambridge, MA

Levine J, Humphreys D (2003) Learning action strategies for planning domains using genetic programming. In: Raidl GR, Meyer JA, Middendorf M, Cagnoni S, Cardalda JJR, Corne D, Gottlieb J, Guillot A, Hart E, Johnson CG, Marchiori E (eds) EvoWorkshops. Lecture notes in computer science, vol 2611. Springer, New York, pp 684–695

Levine J, Westerberg H, Galea M, Humphreys D (2009) Evolutionary-based learning of generalised policies for AI planning domains. In: Rothlauf F (ed) Proceedings of the 11th annual conference on genetic and evolutionary computation (GECCO 2009). ACM, New York, pp 1195–1202

Mandal M, Breaker RR (2004) Gene regulation by riboswitches. Cell 6:451–463

Mathews DH, Turner DH (2002) Dynalign: an algorithm for finding the secondary structure common to two RNA sequences. J Mol Biol 317(2):191–203

Milo N, Zakov S, Katzenelson E, Bachmat E, Dinitz Y, Ziv-Ukelson M (2013) Unrooted unordered homeomorphic subtree alignment of rna trees. Algorithms Mol Biol 8(1):13

Milo N, Yogev S, Ziv-Ukelson M (2014) Stemsearch: Rna search tool based on stem identification and indexing. Methods

Mitchell TM (1999) Machine learning and data mining. Commun ACM 42(11):30–36

Oltean M (2005) Evolving evolutionary algorithms using linear genetic programming. Evol Comput 13(3):387–410. http://dx.doi.org/10.1162/1063656054794815

Pederson J, Bejerano G, Siepel A, Rosenbloom K, Lindblad-Toh K, Lander E, Kent J, Miller W, Haussler D (2006) Identification and classification of conserved RNA secondary structures in the human genome. PLOS Comput Biol 2:e33

Samadi M, Felner A, Schaeffer J (2008) Learning from multiple heuristics. In: Fox D, Gomes CP (eds) Proceedings of the twenty-third AAAI conference on artificial intelligence (AAAI 2008), AAAI Press, pp 357–362

Sczyrba A, Kruger J, Mersch H, Kurtz S, Giegerich R (2003) RNA-related tools on the bielefeld bioinformatics server. Nucleic Acids Res 31(13):3767

Siebert S, Backofen R (2005) MARNA: multiple alignment and consensus structure prediction of RNAs based on sequence structure comparisons. Bioinformatics 21(16):3352–3359

Staple DW, Butcher SE (2005) Pseudoknots: RNA structures with diverse functions. PLoS Biol 3(6):e213

Thompson J, Higgins D, Gibson T (1994) CLUSTALW: improving the sensitivity of progressive multiple sequence alignment through sequence weighting, position-specific gap penalties and weight matrix choice. Nucleic Acids Res 22(22):4673

Torarinsson E, Havgaard JH, Gorodkin J (2007) Multiple structural alignment and clustering of RNA sequences. Bioinformatics 23(8):926–932

Wang Z, Zhang K (2001) Alignment between two RNA structures. Lecture notes in computer science. Springer, Berlin, pp 690–702

Washietl S, Hofacker I (2004) Consensus folding of aligned sequences as a new measure for the detection of functional RNAs by comparative genomics. J Mol Biol 342:19–30

Will S, Reiche K, Hofacker IL, Stadler PF, Backofen R (2007) Inferring non-coding RNA families and classes by means of genome-scale structure-based clustering. PLOS Comput Biol 3(4):e65

Yoon SW, Fern A, Givan R (2008) Learning control knowledge for forward search planning. J Mach Learn Res 9:683–718. http://doi.acm.org/10.1145/1390681.1390705

Kaizen Programming for Feature Construction for Classification

Vinícius Veloso de Melo and Wolfgang Banzhaf

Abstract A data set for classification is commonly composed of a set of features defining the data space representation and one attribute corresponding to the instances' class. A classification tool has to discover how to separate classes based on features, but the discovery of useful knowledge may be hampered by inadequate or insufficient features. Pre-processing steps for the automatic construction of new high-level features proposed to discover hidden relationships among features and to improve classification quality. Here we present a new tool for high-level feature construction: Kaizen Programming. This tool can construct many complementary/dependent high-level features simultaneously. We show that our approach outperforms related methods on well-known binary-class medical data sets using a decision-tree classifier, achieving greater accuracy and smaller trees.

Keywords Kaizen programming • Genetic programming • Classification • Decision-tree

1 Introduction

The objective of a classification algorithm is to predict the class (label) of a record given the values of its attributes. In order to do that, it employs knowledge obtained from a tagged data set, composed of pre-classified records. The information contained in the attribute set (also known as feature set) and in the labels is used to build a model able to accurately differentiate the classes present in the data. This

V.V. de Melo (✉)
Department of Computer Science, Memorial University of Newfoundland,
St. John's, NL, Canada A1B 3X5

Institute of Science and Technology, Federal University of São
Paulo – UNIFESP, São Paulo, Brazil
e-mail: vinicius.melo@unifesp.br

W. Banzhaf
Department of Computer Science, Memorial University of Newfoundland,
St. John's, NL, Canada A1B 3X5

© Springer International Publishing Switzerland 2016
R. Riolo et al. (eds.), *Genetic Programming Theory and Practice XIII*,
Genetic and Evolutionary Computation, DOI 10.1007/978-3-319-34223-8_3

model will then be tested using unseen data which have labels (for quality assurance) and other data without labels (for use the classifier).

The classification process can therefore be seen to consist of two phases. The first phase, which corresponds to model building or training, employs a data set of n records with known labels for each record. The model has to correctly identify the class $(y_i, i = 1, \ldots, n)$ of each record $x_{i,j}, j = 1, \ldots, m$, where m is the number of features. The second phase consists of using this classifier to predict classes of unknown records that were not employed in the training phase. Obviously, to evaluate the classifier's performance in the second phase, the test records must have a known class, which is not used in the prediction but is used for comparison with the class predicted by the model.

The method proposed in this paper aims at an improvement of predictive quality by discovering useful knowledge from data in the pre-processing stage. Such extracted knowledge is inserted into the data set in the form of new attributes and can be used subsequently by the classifier to build new models. This strategy is known as feature construction or feature generation (Liu and Motoda 1998), which can also be employed for dimensionality reduction (Guo et al. 2008).

While many feature construction methods are deterministic (Schölkopf et al. 1997; Nguyen and Rocke 2004; Jolliffe 2005), stochastic approaches have also been proposed (Guo et al. 2008; Neshatian et al. 2012; Wu and Banzhaf 2011). Deterministic methods rely on greedy heuristics that are supposed to work on any kind of data, but have been shown to not always effective (Wolpert and Macready 1997). Stochastic approaches are more flexible in this aspect: They can generate and evaluate non-linear features that would be discarded by deterministic methods. It is easy to see that, by being more rigid, greedy deterministic methods tend to be much faster but less capable of exploring the search-space, while stochastic methods will be slower but may generate better features.

The method reported in this chapter combines a stochastic and a deterministic method into a hybrid method. Its stochastic part performs knowledge extraction to generate high-level features, and its deterministic part builds a classification model on top of that. In this work we aim at a grey-box classifier, which is a human-readable model that may not be fully understandable ("gray" instead of "white") because some formulas in the new features could be complex and opaque, though clearer than results produced by black-box approaches such as Artificial Neural Networks.

In the present contribution we employ a collaborative approach to search for high-quality features. There are many aspects that differentiate our method from others found in the literature that use a team-based approach, see, for instance (Brameier and Banzhaf 2001). Those differences will be explained later. For now, to say that our approach evolves a set of features instead of evolving individual features to be used as an ensemble, as commonly explored in the literature.

In order to have a good ensemble, it is important that all classifiers have high predictive quality. Therefore, it is reasonable to suppose that there is a high chance of having very similar classifiers that do not augment each other. The methodology of the present study builds a single classifier whose features are likely to be complementary to each other. It is similar in behavior to PCA (Jolliffe 2005), where the features generated show a decreasing degree of variance the data. The difference is, however, that our method discovers non-linear features using arbitrary formulas.

In order to perform an efficient search, our approach is based on the Kaizen methodology (Imai 1986), that will be briefly introduced below. The main idea of Kaizen is the continuous improvement of a process through the PDCA (Plan-Do-Check-Act, Gitlow et al. 1989) cycle, generating a new solution based on the knowledge obtained in previous cycles. This new solution can be divided to conquer, allowing individual analysis and improvement of each part. Therefore, a solution is actually composed of partial solutions. Our approach, called Kaizen Programming (KP, de Melo 2014) is an implementation of the PDCA cycle. KP can generate a feature set, build and evaluate the model, extract the importance of each feature from the set, and evolve useful attributes to extract high-quality knowledge from the training data.

The rest of this chapter is organized as follows. Section 2 introduces the concept of Feature Construction, Section 3 describes related algorithms for feature constructions that were used for comparison in this work. Section 4 presents Kaizen Programming applied to feature construction. Computational experiments are presented in Sect. 5 with Sect. 6 providing some conclusions.

2 Feature Construction

Feature construction (Liu and Motoda 1998; Guyon et al. 2006; Kantardzic 2011) is a process employed to discover useful knowledge regarding hidden relationships among features in a set of data. The newly constructed features can then be either used alone or to augment the existing data set. When used alone, the new features may be smaller in number than the original feature set, acting as a dimensionality reduction method; thus named Feature Extraction (Liu and Motoda 1998; Guyon et al. 2006; Kantardzic 2011). It is expected that the new features perform a better separation of classes, facilitating the data mining task.

As pointed out by Freitas (2008), when compared to feature selection, the construction of new useful attributes can be a much more difficult task. This can be explained by the fact that feature selection is a binary combination problem (a feature is *selected* or *not selected*, giving 2^k possibilities, where k is the number of features), while construction is a multi-valued combination task because the new feature is a function composition.

Many methods have been proposed in the last decades to perform feature construction for dimensionality reduction. Deterministic approaches such as Principal Component Analysis (PCA, Jolliffe 2005), and Partial Least Squares (PLS, Nguyen and Rocke 2004) are usually fast and able to find useful features. However, these techniques do not search for a global optimum. Stochastic techniques such as evolutionary algorithms have been investigated to that end. As examples of distinct methods from the area of Genetic Programming (Banzhaf et al. 1998) we refer to Smith and Bull (2005), Gavrilis et al. (2008), and Drozdz and Kwasnicka (2010).

After generating a feature one must measure its quality. As we are aiming to achieve the best prediction quality, we selected the *wrapper* approach (Miner et al. 2009), where a predictive model (a classifier, for instance) is built using the input feature set, and the prediction results are compared to the expected results. The percentage of correct predictions made on the test set is used as a quality measure. Since the wrapper does not provide feature importance, it is necessary to investigate how the classifier used the features to build the model in. This information about feature importance is then used to efficiently guide the search.

Our models are built using a random sample from the data set (the training set), and tested using a distinct sample from the same data set. Wrappers may require a large computational effort, but their use tends to result in high-quality features that will be tuned to the specific classifier. Finding this best solution implies a global search and a stochastic algorithm is a reasonable choice for this task. The next Section presents some related work.

3   Evolutionary Algorithms for Feature Construction

While there are many using evolutionary algorithms for feature construction, here we briefly introduce only those techniques that are used for comparison in the experimental section.

GPMFC+CART (Neshatian et al. 2012): This technique is a GP-based system for construction of multiple features for classification problems. It uses a filter approach instead of a wrapper, to evaluate the quality of the constructed features during the evolution. The multiple features are sequentially constructed, one by one, for each class of the dataset, maximizing the purity of class intervals. After evolution, the features were tested using the CART decision-tree technique.

MLGP (Wu and Banzhaf 2011): In this contribution, the multilevel selection framework (Wu and Banzhaf 2010) served as inspiration to the development of a multilevel genetic programming approach (MLGP). Multilevel selection tries to encourage cooperation among individuals based on biological group selection theory. The authors developed a cooperation operator in order to build solutions hierarchically. The fitness of a group (or individual) is calculated through direct evaluation, without external classifiers. Therefore, the individual, or group, is used to classify the data.

GP-EM (Guo et al. 2010): This method uses GP to generate a single feature, and an expectation maximization algorithm (EM) to model the new features as a Gaussian mixture (GM). The objective is to evolve features that better separate the classes when modeled as GM. The fitness measure, in this case, considers both the within-class scatter and the between-class scatter values.

GP+C4.5 (Neshatian et al. 2007): In this contribution, classical GP is used to evolve multiple features, and a class-dispersion and entropy-based measure is employed to calculate a feature's quality. A feature is independently constructed for each class in the dataset. Therefore, the distribution of classes in a particular feature must be well separated. After evolving the features, experiments were performed using the well-known C4.5 classifier.

GP+CART (Muharram and Smith 2004): These authors employ two distinct fitness measures to evolve features using GP: Information Gain and Gini Index. The constructed feature is assumed to be a node in a decision-tree, and fitness is calculated using the result of a split in that node. A single feature is evolved in each GP run, and four classifiers are tested on the features. We selected the results obtained by CART to compare to the results herein.

In the next section we describe KP and our proposal for feature construction.

4 Kaizen Programming Applied to Feature Construction

Kaizen Programming (KP), proposed by de Melo (2014), is a novel tool inspired by the concepts of the Kaizen method (Imai 1986). KP is a computational implementation of a Kaizen event with the Plan-Do-Check-Act (PDCA) methodology employed to guide a process continuous improvement. However, KP is an *abstraction* of the main components of PDCA.

Compared with classical GP, KP follows a different method for the automated design of algorithms. KP individuals are not complete solutions, only parts of solutions that have to combine together. As a result, evolution becomes a collaborative approach with the expectation that more than one partial solution is improved to help other partial solutions.

In KP, a team of experts is formed to propose ideas to solve a problem, which then are joined to become a solution. The quality of a solution measures how well it solves the problem, and the quality of an idea quantifies its contribution/importance to the solution. KP first builds a model, and then calculates the importance of each feature. Therefore, different from general GP and other evolutionary algorithms that perform trial-and-error search guided by natural selection, in KP can determine, exactly which parts of the solution should be removed or improved because they were important to the method that built the model. Consequently, the experts contribute by providing better ideas in each cycle. This results in a reduction in bloat, population size, and number of function evaluations. A further difference to other team-based approaches is that the team in KP is a set of agents (data structure + procedures), while for other methods a team is a set of solutions (individuals).

Conceptually, the *knowledge* acquired during the search is shared with the team to improve everyone's ideas. Thus, there is a single set of ideas accessible to all experts, not multiple populations. Not all experts may provide useful contributions all the time that is, the search mechanism does not guarantee that every cycle will give a better solution. However, it is expected that better ideas are generated over the cycles. A brief explanation of the PDCA cycle is presented next.

REPEAT

> **PLAN**: assuming the current ideas (called *standard*) the team performs a *brainstorming*, and each expert proposes one or more ideas to solve part of the problem;
> **DO**: the standard and new ideas are applied (executed/parsed/evaluated/ calculated) to the problem and put together to become a complete solution;
> **CHECK**: evaluate the proposed solution, then each single idea (considering the standard and the new ones) is analyzed and its contribution to solve the problem is measured. Create a new solution using only the important ideas and measure its quality;
> **ACT**: if the solution quality has improved, then the standard is updated, which is presented to the team along with each contribution, improving the knowledge of the problem. Create another kaizen event with a new team if the current one doesn't improve the standard after a certain number of cycles;

WHILE target not achieved

In this chapter, KP is employed to perform high-level feature construction to improve prediction quality of a particular classifier. Various features can be generated at the same time, being improved over PDCA cycles. As opposed to what happens in traditional approaches, in KP those features are dependent on each other, therefore the result is a feature set for a single model, not an ensemble.

4.1   Implementation

Algorithm 1 presents the pseudo-code of the KP method implemented for this contribution. The experts work on a tree-based representation, i.e., as a traditional GP, and may perform only recombination (crossover), only variation (mutation), or both.

The ideas proposed by the experts are non-linear combinations of the original features (formulas) using the terminals and non-terminals defined by the user. The ideas are randomly selected for improvement (there is no tournament) as all of them are supposed to be important. To facilitate implementation, we assumed that the number of experts is the same as the number of features to be constructed, but they are actually distinct parameters. The Expansion Factor to increase the size of the team is a mechanism that may be used when stagnation is detected.

The method selected for building the model was the Classification and Regression Tree algorithm (CART, Breiman et al. 1984). Also, our CART implementation (Pedregosa et al. 2011) provides the Gini Importance (Breiman 2001) for each feature of the dataset, which is used as the importance measure. Thus, one may notice that CART must be used twice: first with all features to measure

---

Algorithm 1 Pseudo-code of Kaizen Programming for feature construction

---

1. **Read** the dataset and set n as the number of instances
2. **Set** *CurrentStandardQuality* $\leftarrow$ 0, *MaxStagnated*, *Stagnated* $\leftarrow$ 0, Size of the Team (s_t), number of New Ideas per Expert (*NIE*), Expansion Factor (*EF*), $w \leftarrow s_t * NIE$
3. **Define** the target and set it as not achieved
4. **Generate** s_t initial random ideas as *CurrentStandard*
5. **Apply** the *CurrentStandard* (calculate the results from the expressions) and create the feature set STD_{n,s_t}
6. *BestStandard* $\leftarrow$ *CurrentStandard*
7. *BestStandardQuality* $\leftarrow$ *CurrentStandardQuality* on k-fold cross-validation
8. **Do**

    a. **Generate**, via GP operators, the *TrialIdeas*, which are *NIE* variations (ideas) of the *CurrentStandard* through multiple crossover and mutation. Even the worst idea from *CurrentStandard* might have offspring

    b. **Apply** each new idea, resulting in the $TRIAL_{n,w}$ feature set

    c. **Create** the expanded feature set F_{n,s_t+w} containing $TRIAL_{n,w}$ and STD_{n,s_t}

    d. **Create** new k stratified folds from F to reduce bias in the search

    e. **For** each fold

        i. **Induce** a decision tree via CART

        ii. **Calculate** the array *FoldImportances* as the importance of each feature from F using *Gini Importance*

    f. **End For**

    g. **Set** *TrialImportances* as the average of all *FoldImportances*

    h. *MostImportantTrialIdeas* is the subset of the s_t most important *TrialIdeas* (considering *TrialImportances*)

    i. **Create** $MITI_{n,s_t}$ as a subset of F, and calculate *MostImportantTrialIdeasQuality* using the current k-folds

    j. **If** *MostImportantTrialIdeasQuality* is better than *CurrentStandardQuality* then

        i. *CurrentStandard* $\leftarrow$ *MostImportantTrialIdeas*

        ii. *CurrentStandardQuality* $\leftarrow$ *MostImportantTrialIdeasQuality*

        iii. $STD_{n,s_t} \leftarrow MITI_{n,s_t}$

        iv. **If** *CurrentStandardQuality* is better than *BestStandardQuality* then

            A. *BestStandard* $\leftarrow$ *CurrentStandard*

            B. *BestStandardQuality* $\leftarrow$ *CurrentStandardQuality*

        v. **End If**

    k. **Else**

        i. *Stagnated* $\leftarrow$ *Stagnated* + 1

    l. **End If**

    m. **If** *Stagnated* > *MaxStagnation* then

        i. *Stagnated* $\leftarrow$ 0

        ii. $s_t \leftarrow s_t + \lceil s_t * EF \rceil$ to increase the team of experts' size

        iii. **Generate** s_t initial random ideas as *CurrentStandard*

        iv. **Apply** the *CurrentStandard* (calculate the results from the expressions) and create STD_{n,s_t}

        v. **Calculate** *CurrentStandardQuality* on k-fold cross-validation

    n. **End If**

9. **While** target is not achieved
10. **Return** *BestStandard*, *BestStandardQuality*

---

their importance, and then with the reduced feature set to measure the actual solution quality. Therefore there is an expansion of the feature set, followed by feature selection. Finally, to reduce the risk of overfitting we used cross-validation in the training.

5 Experiments

This section presents our experiments performed to evaluate KP for classification. KP was tested using publicly available two-class medical datasets from the UCI online repository (Lichman 2013). Some characteristics of the datasets are presented in Table 1. The datasets were chosen after selecting papers from literature that will be used for comparison.

5.1 Pre-processing

Given that KP generates mathematical expressions using features from the dataset, it is necessary to prepare the data. The Weka machine learning tool (Hall et al. 2009) was used to replace missing values with the means from the training data, instead of removing incomplete instances. No other transformation, normalization, or standardization was performed on the data.

5.2 Computational Environment

KP was implemented in the Python programming language (version 2.7.6), using GP from DEAP (Distributed Evolutionary Algorithms in Python) library (version 1.0.1), and scikit-learn library (version 0.14.2) for CART. To evaluate the features discovered by KP, tests were performed using CART in Weka (version 3.6.11) running on Java (version 1.7.0_55) via OpenJDK Runtime Environment (IcedTea version 2.4.7). The experiments were executed on an Intel i7 920 desktop, with 6Gb of RAM, Archbang Linux (kernel version 3.14.5-1), GCC (version 4.9.0 20140521).

Table 1 Summary of the two-class datasets employed in the experiments

Dataset	Continuous attributes	Instances
Breast-w (Winsconsin)	9	699
Diabetes (PIMA)	8	768
Liver-disorders (BUPA)	6	345
Parkinson	22	195

5.3 Organization of the Experiments

During the discovery phase (training), a k-fold stratified cross-validation was performed to calculate both the importance of ideas and the solution quality of selected ideas. It is important to be clear that KP did not evolve features for a specific k-fold configuration, because every time the objective function was called k new stratified folds were generated.

For each dataset of Table 1, KP was run 32 independent times with a different random seed.[1] All runs used the configuration shown in Tables 2, 3, and 4. KP was configured to search for the same number of ideas (10 new features), independently of the number of features in the original dataset. However, not all may be used in the final classifier.

In the expert configuration, GP evolutionary operators, $pdiv$, $plog$, and $psqrt$ are protected versions of these operations. $pdiv(a, b)$ returns zero whenever b is zero; $plog(a)$ returns zero whenever the a is zero, and $log(abs(a))$ otherwise; $psqrt(a)$ returns $1e100$ if $a \leq 0$; and $hypot(a, b) = sqrt(a * a + b * b)$.

Since the CART implementation in scikit-learn is not exactly the same as in Weka, it was necessary to use two parameters to achieve greater similarity between the results of different implementations: maximum tree-depth and minimum objects in the leaf node. Features were then tested, in the second phase, with distinct configurations of the CART method (in Weka), which also performed the statistical analysis. This experiment was to evaluate the decision-tree's performance using the original feature set (O), the new feature set (N) discovered by KP, and the

Table 2 KP and CART configuration

Parameter	Value
Initial experts (s_t)	10
Initial ideas generator	GP ramped half-half
Initial ideas max. depth	2
New ideas per expert (NIE)	5
Cycles	2000
Stagnation	2.5 % of the cycles
Factor (EF) to increase experts	0, disabled
Independent runs	32
Model builder (decision-tree)	CART
CART Max. depth	5
CART Min. instances at a leaf	10
k (folds)	10
Solution quality/fitness	*Accuracy*
Idea importance	*Gini Importance*

[1] Thirty-two runs were performed because it is a multiple of 8, and the runs were done in parallel on a quad-core machine with hyper-threading, so we employed all available processing units.

Table 3 Experts configuration (GP operators)

Parameter	Value
Crossover probability	0.2
Idea combinator/crossover operator	One-point
Mutation probability	1.0
Idea improver/mutation operator	GP subtree replacement
Max. depth	10
Non-terminals	$+, -, \times, pdiv(a, b), plog, psqrt, neg, cos, sin, tan,$ $tanh, square, cube, sigmoid, hypot(a, b), max(a, b),$ $min(a, b), avg2(a, b), avg3(a, b, c)$
Terminals	$x_i, i = 1, \ldots, nf$ (features of the original dataset)

Table 4 CART configuration in Weka for the Test phase

Config. name	Min.Number.Obj	Prune	Use OneSE rule
CART_1	2	No	No
CART_2	2	Yes	No
CART_3	2	Yes	Yes
CART_4	10	No	No
CART_5	10	Yes	No
CART_6	10	Yes	Yes

The other parameters were the default values

combination of new and original feature sets (*NO*). In Weka, CART was configured in six different ways (see Table 4) to verify the influence of the pruning mechanism.

The second analysis is the comparison of the best results obtained by CART experiments versus other feature construction techniques, mainly using Genetic Programming, whose results are reported in the literature. We selected only those that performed ten-fold cross-validation.

5.4 Method of Analysis

The results presented here are only from the test phase. Given that KP was run 32 times on each dataset, we have 32 new feature sets for each of them. A CART decision-tree was induced for each feature set using tenfold cross-validation. Therefore, the original dataset gives 10 results, while each new feature set gives $32 \times 10 = 320$ results.

The evaluated measures were Accuracy, Weighted *F*-Measure, and Tree size. Accuracy considers the whole dataset, while the Weighted *F*-Measure is the sum of all *F*-measures, each weighted according to the number of instances with that particular class label. Tree size is used to evaluate the complexity of the final solution; however, it does not take into consideration the complexity of a feature.

The relevance of this information can be decided by the user when defining the maximum tree-depth used by KP when generating new ideas (features).

In order to verify if there are differences between the feature sets (O versus N, and O versus NO), we executed Welch's t-test at a significance level $\alpha = 0.05$. If the new features result in statistically different means, a mark '*' is inserted after the standard-deviation in the tables showing the results.

5.5 Evaluation of the Discovered Features

For each dataset investigated here, one has a table with a short descriptive analysis (mean and standard-deviation) of the results for each CART configuration and feature sets, with the significant differences (via Welch's test) marked when necessary. The discussion on the results is as follows.

For the Breast Cancer dataset, one can see the short descriptive analysis in Table 5. Accuracy when using either the New features (N) or the combination of New and Original features (NO) improved significantly, as shown by the symbol '*'. It is interesting to notice that for both configurations CART_5 and CART_6, the accuracies using N and NO were identical. This suggests that CART used only the new features from the NO dataset; therefore, the Original features (O) were not very useful anymore. This hypothesis gets stronger when configurations CART_2, CART_3, and CART_4 are analyzed, in which the mean accuracy of using N is bigger than using NO. The highest mean accuracy was achieved using a minimum of 2 instances for leaf and the pruning mechanism without the OneSE rule (CART_2).

The second classification quality measure is the Weighted F-Measure, which considers the correct classification of each class separately. Again, all CART configurations presented statistically better results when using N. For unbalanced datasets, where one class has considerably more instances than the other, these two measures may not have the same statistical interpretation.

The third measure is the tree size. Given that N is more representative than O, a significant reduction is expected. As shown in the corresponding table, this reduction was bigger than 50 % for CART_1, CART_2, and CART_3, all of them using minimum number of leaves set 2. A relevant comparison can be made between the results of CART_1 and CART_3: there was an increment in the accuracy (from 93.7 to 97.28 %) and a reduction in the tree size (from 41 to 4.41). Consequently, by the results present in this table, the features discovered by KP for the Wisconsin Breast Cancer dataset helped CART in finding better and smaller trees.

Regarding the PIMA diabetes dataset (Table 6), the lowest accuracy occurred using CART_1 on the O dataset, while the highest accuracy was obtained with CART_2 on the N dataset. All N datasets improved over O and were also better than all NO. Similar behavior is present in the Weighted F-Measure results. A posterior application of feature selection on NO could help improving the accuracy. With respect to the trees sizes, large reductions can be seen from CART_1 to CART_2, with corresponding increase in Accuracy and Weighted F-Measure. However, in

Table 5 Short descriptive analysis (mean and standard-deviation) for the breast-w dataset

Metric	Feat	CART_1	CART_2	CART_3	CART_4	CART_5	CART_6
Accuracy	O	93.70 (2.80)	94.42 (3.53)	94.27 (3.93)	93.99 (4.03)	94.13 (3.96)	93.56 (4.73)
Accuracy	N	97.21 (2.16)*	97.43 (2.03)*	97.28 (2.02)*	97.44 (1.99)*	97.12 (2.66)*	97.04 (2.64)*
Accuracy	NO	97.22 (2.16)*	97.38 (2.07)*	97.25 (2.05)*	97.43 (1.99)*	97.12 (2.66)*	97.04 (2.64)*
W. F-Meas.	O	0.94 (0.028)	0.94 (0.035)	0.94 (0.039)	0.94 (0.041)	0.94 (0.04)	0.94 (0.047)
W. F-Meas.	N	0.97 (0.021)*	0.97 (0.02)*	0.97 (0.02)*	0.97 (0.02)*	0.97 (0.032)*	0.97 (0.032)*
W. F-Meas.	NO	0.97 (0.022)*	0.97 (0.02)*	0.97 (0.02)*	0.97 (0.02)*	0.97 (0.032)*	0.97 (0.032)*
Tree size	O	41.00 (3.53)	16.80 (7.33)	9.00 (5.08)	15.40 (1.26)	5.60 (0.97)	5.20 (1.14)
Tree size	N	18.89 (4.19)*	7.02 (3.58)*	4.41 (2.35)*	8.00 (2.80)*	3.76 (1.58)*	3.18 (0.86)*
Tree size	NO	18.34 (4.22)*	6.74 (3.36)*	4.25 (2.21)*	7.96 (2.78)*	3.76 (1.58)*	3.18 (0.86)*

Table 6 Short descriptive analysis (mean and standard-deviation) for the diabetes dataset

Metric	Feat	CART_1	CART_2	CART_3	CART_4	CART_5	CART_6
Accuracy	O	71.75 (3.39)	75.13 (4.09)	74.36 (4.07)	75.26 (6.20)	75.27 (3.69)	74.09 (4.43)
Accuracy	N	75.54 (4.51)*	79.65 (4.65)*	78.58 (4.73)*	79.48 (4.59)*	79.45 (4.79)*	78.36 (4.79)*
Accuracy	NO	74.62 (4.70)	79.17 (4.89)*	78.19 (4.83)*	79.02 (4.53)*	79.08 (4.67)*	78.05 (4.76)*
W. F-Meas.	O	0.71 (0.035)	0.74 (0.048)	0.73 (0.043)	0.75 (0.061)	0.74 (0.041)	0.73 (0.044)
W. F-Meas.	N	0.75 (0.046)*	0.79 (0.049)*	0.78 (0.051)*	0.79 (0.047)*	0.79 (0.05)*	0.78 (0.052)*
W. F-Meas.	NO	0.75 (0.047)	0.79 (0.052)*	0.78 (0.052)*	0.79 (0.047)*	0.79 (0.05)*	0.77 (0.051)*
Tree size	O	131.80 (11.93)	16.20 (13.54)	4.40 (1.35)	27.80 (5.98)	8.60 (8.37)	4.20 (1.40)
Tree size	N	105.51 (14.74)*	15.27 (7.87)	9.19 (4.67)*	26.66 (5.69)	13.84 (5.33)*	8.57 (4.47)*
Tree size	NO	104.69 (14.45)*	14.36 (8.02)	8.01 (4.51)*	26.18 (5.13)	13.38 (5.81)*	7.50 (4.15)

contrast to what happened to the breast-w dataset, in this case the sizes were bigger when N and NO were used by CART_3, CART_5, and CART_6. We are still investigating the results to propose a reasonable explanation for this issue.

In the BUPA liver-disorders dataset (Table 7) both Accuracy and Weighted F-Measure improved more than $11.94\% = (75\% - 67\%)/67\%$ when the discovered features were employed. As will be seen in the comparison with results from the literature, this improvement is very relevant. Finally, the increase in the trees sizes is present for the same CART configurations as in the previous dataset. Nevertheless, smaller trees showed similar or better quality than the bigger ones.

The last dataset contains information of Parkinson's disease. The most noticeable characteristic in Table 8 is that even though both the mean Accuracy and Weighted F-Measure improved, the standard-deviations were large, reflecting non-significant differences for configurations CART_4, CART_5, and CART_6, that had as a termination criterion a minimum of 10 instances per leaf. Therefore, it was better to let the tree grow deeper and prune it afterwards, taking the risk of overfitting. This means that, for this dataset, for a significant number of times KP did not discover features capable of reducing entropy in the leaf nodes. A possible explanation is that, as shown in Table 1, this dataset has not only more attributes than the other three datasets, but also fewer instances. Therefore, either a longer run would be necessary or one would need more than 10 features. Nevertheless, the new features led to an increase in mean Accuracy from 87.68 % (best solution using O) to 93.85 % (best solution using N or NO).

5.6 Comparison Against Other Feature Construction Techniques

In this section, KP's results are compared with those from the literature. In order to have a fairer comparison, we selected only works using GP (or a similar technique) to evolve features, with ten-fold cross-validation in the test phase. The comparison is performed with techniques presented in the literature review: GPMFC+CART, MLGP, GP-EM, GP+C4.5, and GP+CART. The results on other methods were taken from the other authors' original works.

From each dataset in the previous section, we have selected the highest mean Accuracy among the CART configurations (see Table 9). Not all datasets used in this work were found in other papers.

As one can see, the features discovered by KP led to more accurate classifiers than all the other feature construction techniques. An important characteristic the number of feature sets created by the techniques. For KP, two feature sets have to be tested at each generation: the first one using the current ideas (features) and the new ideas simultaneously to calculate the importance of each idea; the second one using only the s_t most important ideas to finally calculate the solution quality. Given that KP was run for 2000 cycles, 4000 feature sets were generated in the

Table 7 Short descriptive analysis (mean and standard-deviation) for the liver-disorders dataset

Metric	Feat	CART_1	CART_2	CART_3	CART_4	CART_5	CART_6
Accuracy	O	67.54 (8.32)	67.57 (7.94)	65.85 (8.88)	68.66 (6.37)	65.26 (7.59)	65.25 (8.22)
Accuracy	N	76.23 (7.55)*	78.86 (7.27)*	77.10 (7.24)*	78.80 (7.51)*	77.64 (7.08)*	74.95 (7.15)*
Accuracy	NO	75.71 (7.69)*	78.34 (7.32)*	76.50 (7.24)*	78.28 (7.42)*	77.18 (7.23)*	74.77 (7.18)*
W. F-Meas.	O	0.67 (0.083)	0.66 (0.087)	0.64 (0.089)	0.68 (0.065)	0.64 (0.081)	0.62 (0.10)
W. F-Meas.	N	0.76 (0.076)*	0.79 (0.073)*	0.77 (0.075)*	0.79 (0.076)*	0.77 (0.072)*	0.74 (0.075)*
W. F-Meas.	NO	0.75 (0.077)*	0.78 (0.074)*	0.76 (0.075)*	0.78 (0.075)*	0.77 (0.074)*	0.74 (0.075)*
Tree size	O	79.40 (12.64)	20.80 (14.92)	8.00 (5.83)	22.20 (3.43)	10.00 (4.55)	5.80 (3.16)
Tree size	N	53.48 (9.96)*	17.76 (7.48)	11.43 (4.20)*	18.16 (4.07)*	12.62 (4.17)	8.57 (4.14)
Tree size	NO	50.71 (9.29)*	17.26 (8.02)	10.94 (4.53)	17.62 (3.87)*	12.21 (4.39)	8.18 (4.16)

Table 8 Short descriptive analysis (mean and standard-deviation) for the parkinsons dataset

Metric	Feat	CART_1	CART_2	CART_3	CART_4	CART_5	CART_6
Accuracy	O	87.68 (4.27)	85.66 (4.57)	87.26 (8.23)	86.71 (7.12)	86.21 (8.21)	84.71 (8.54)
Accuracy	N	93.32 (5.34)*	93.85 (5.53)*	93.79 (5.64)*	92.25 (6.68)*	91.20 (7.03)	90.40 (7.11)
Accuracy	NO	92.80 (5.64)*	93.17 (5.85)*	93.03 (6.17)*	91.71 (6.86)	90.60 (7.37)	89.57 (7.64)
W. F-Meas.	O	0.88 (0.041)	0.86 (0.047)	0.87 (0.084)	0.86 (0.073)	0.86 (0.08)	0.84 (0.089)
W. F-Meas.	N	0.93 (0.054)*	0.94 (0.056)*	0.94 (0.059)*	0.92 (0.069)*	0.91 (0.074)	0.90 (0.076)
W. F-Meas.	NO	0.93 (0.057)*	0.93 (0.06)*	0.93 (0.065)*	0.91 (0.072)	0.90 (0.078)	0.89 (0.081)
Tree size	O	17.60 (4.43)	10.80 (3.46)	5.40 (2.80)	8.40 (1.35)	5.80 (2.70)	3.80 (1.69)
Tree size	N	14.20 (2.85)*	8.65 (2.85)	7.26 (2.12)*	7.44 (1.80)	5.83 (1.85)	5.02 (1.82)
Tree size	NO	13.52 (3.25)*	8.54 (2.76)*	7.12 (2.16)*	7.28 (1.69)	5.72 (1.86)	4.74 (1.83)

Table 9 Comparison of mean accuracy among feature extraction techniques that use GP

Dataset	KP+CART	GPMFC+CART	MLGP	GP-EM	GP+C4.5	GP+CART
Breast-w	97.44	96.3**	96.8	–	97.2**	–
Diabetes	79.65	–	71.6	–	75.4	–
Liver-disorders	78.86	67.68	67.5	–	70.4	69.71
Parkinsons	93.85	–	–	93.12	–	–
Feature sets	4000	100,000	600,000	11,200	18,000	60,000

Symbol '**' means a reduction in the number of instances due to missing values, and "–" means Not Available

process. Even though a ten-fold cross-validation approach was used in the training phase, the features were the same for all folds. Because the features in KP are partial solutions, they cannot be evaluated separately.

On the other hand, for the other techniques from Table 9 a single individual is a solution to the problem thus they employed more feature sets. As most techniques evolve a single expression per solution/class, more runs are necessary to have a set of features, while KP can evolve many complementary features at the same time. For them, we calculated the number of feature sets as Population size × number of generations × number of features generated. An interesting conjecture is that in order to achieve a performance close to that shown by KP, other techniques may need a more complex formula, while KP may generate a set of smaller/simpler formulas allowing for a posterior feature selection procedure, if desired by the user.

6 Conclusions

This chapter presented Kaizen Programming (KP) as a technique to perform high-level feature construction. KP evolves partial solutions that complement each other to solve a problem, instead of producing individuals that encode complete solutions.

Here, KP employed tree-based evolutionary operators to generate ideas (new features for the dataset) and the CART decision-tree technique for the wrapper approach. The gini impurity used by CART as split criterion is used to calculate the importance of each feature, translating into the importance of each partial solution in KP. The quality of complete solutions was calculated using accuracy in a tenfold stratified cross-validation scheme.

Four widely studied datasets were used to evaluate KP, and tests were performed on six distinct CART configurations. Comparisons among different configurations were made in terms of mean and standard deviation of accuracy, weighted f-measure, and tree-size. A hypothesis test was performed to compare the mean performance when using the new features, and the new and original features together. Results show that the new features with or without the original ones, improved performance and reduced tree-sizes significantly.

The second comparison was against five related approaches from the literature. All those approaches employ genetic programming to construct features from the original dataset and test them using well-known classifiers. It was found that KP was better than all other approaches, while requiring a fraction of the feature sets generated in other work. KP was not only more accurate, but also much faster.

As future work, a deeper sensitivity analysis will be necessary to verify KP's behavior on distinct configurations in order to be able to differentiate poor from good configurations.

Acknowledgements This paper was supported by the Brazilian Government CNPq (Universal) grant (486950/2013-1) and CAPES (Science without Borders) grant (12180-13-0) to Vinícius Veloso de Melo, and Canada's NSERC Discovery grant RGPIN 283304-2012 to Wolfgang Banzhaf.

References

Banzhaf W, Nordin P, Keller R, Francone F (1998) Genetic programming - an introduction. Morgan Kaufmann, San Francisco

Brameier M, Banzhaf W (2001) Evolving teams of predictors with linear genetic programming. Genet Program Evolvable Mach 2(4):381–407

Breiman L (2001) Random forests. Mach Learn 45(1):5–32

Breiman L, Friedman J, Stone C, Olshen R (1984) Classification and regression trees. The Wadsworth and Brooks-Cole statistics-probability series. Taylor & Francis, London

de Melo VV (2014) Kaizen programming. In: Proceedings of the 2014 conference on genetic and evolutionary computation (GECCO). ACM, New York, pp 895–902

Drozdz K, Kwasnicka H (2010) Feature set reduction by evolutionary selection and construction. In: Agent and multi-agent systems: technologies and applications. Springer, Berlin, Heidelberg, pp 140–149

Freitas AA (2008) A review of evolutionary algorithms for data mining. In: Soft computing for knowledge discovery and data mining, Springer, Berlin, pp 79–111

Gavrilis D, Tsoulos IG, Dermatas E (2008) Selecting and constructing features using grammatical evolution. Pattern Recogn Lett 29(9):1358–1365. doi:10.1016/j.patrec.2008.02.007. http://www.sciencedirect.com/science/article/B6V15-4S01WDH-4/2/aaff3c40c5eca125dfacb 426d88fa177

Gitlow H, Gitlow S, Oppenheim A, Oppenheim R (1989) Tools and methods for the improvement of quality. Irwin series in quantitative analysis for business. Taylor & Francis, London

Guo H, Zhang Q, Nandi AK (2008) Feature extraction and dimensionality reduction by genetic programming based on the fisher criterion. Expert Syst 25(5):444–459

Guo PF, Bhattacharya P, Kharma N (2010) Advances in detecting parkinson's disease. In: Zhang D, Sonka M (eds) Medical biometrics. Lecture notes in computer science, vol 6165. Springer, Berlin, Heidelberg, pp 306–314

Hall M, Frank E, Holmes G, Pfahringer B, Reutemann P, Witten IH (2009) The weka data mining software: an update. SIGKDD Explor Newsl 11(1):10–18. doi:10.1145/1656274.1656278. http://doi.acm.org/10.1145/1656274.1656278

Imai M (1986) Kaizen (Ky'zen), the key to Japan's competitive success. McGraw-Hill, New York

Isabelle G, André E, An introduction to feature extraction. In: Guyon I, Gunn S, Nikravesh M, Zadeh LA (eds) Feature extraction: foundations and applications (Studies in Fuzziness and Soft Computing). Springer, Berlin/Heidelberg, pp 1–25. doi:10.1007/978-3-540-35488-8

Jolliffe I (2005) Principal component analysis. Wiley Online Library

Kantardzic M (2011) Data mining: concepts, models, methods, and algorithms. Wiley, New York

Lichman M (2013) UCI machine learning repository. http://archive.ics.uci.edu/ml

Liu H, Motoda H (1998) Feature extraction, construction and selection: a data mining perspective. Springer, Berlin

Miner G, Nisbet R, Elder IVJ (2009) Handbook of statistical analysis and data mining applications. Academic Press, New York

Muharram MA, Smith GD (2004) Evolutionary feature construction using information gain and gini index. In: Genetic programming, Springer, pp 379–388

Neshatian K, Zhang M, Johnston M (2007) Feature construction and dimension reduction using genetic programming. In: AI 2007: advances in artificial intelligence. Springer, Berlin, pp 160–170

Neshatian K, Zhang M, Andreae P (2012) A filter approach to multiple feature construction for symbolic learning classifiers using genetic programming. Trans Evol Comp 16(5):645–661

Nguyen DV, Rocke DM (2004) On partial least squares dimension reduction for microarray-based classification: a simulation study. Comput Stat Data Anal 46(3):407–425

Pedregosa F, Varoquaux G, Gramfort A, Michel V, Thirion B, Grisel O, Blondel M, Prettenhofer P, Weiss R, Dubourg V, Vanderplas J, Passos A, Cournapeau D, Brucher M, Perrot M, Duchesnay E (2011) Scikit-learn: machine learning in Python. J Mach Learn Res 12:2825–2830

Schölkopf B, Smola A, Müller KR (1997) Kernel principal component analysis. In: Artificial neural networks–ICANN 97. Springer, Berlin, pp 583–588

Smith MG, Bull L (2005) Genetic programming with a genetic algorithm for feature construction and selection. Genet Program Evolvable Mach 6(3):265–281

Wolpert DH, Macready WG (1997) No free lunch theorems for optimization. IEEE Trans Evol Comput 1(1):67–82

Wu SX, Banzhaf W (2010) A hierarchical cooperative evolutionary algorithm. In: Proceedings of the 12th annual conference on genetic and evolutionary computation, GECCO '10. ACM, New York, pp 233–240

Wu SX, Banzhaf W (2011) Rethinking multilevel selection in genetic programming. In: Proceedings of the 13th annual conference on genetic and evolutionary computation, Dublin, pp 1403–1410

GP As If You Meant It: An Exercise for Mindful Practice

William A. Tozier

Abstract In this contribution I present a *kata* called "GP As If You Meant It", aimed at advanced users of genetic programming. Inspired by *code katas* that are popular among software developers, it's an exercise designed to help participants hone their skills through mindful practice. Its intent is to surface certain unquestioned habits common in our field: to make the participants painfully aware of the tacit *justification* for certain GP algorithm design decisions they may otherwise take for granted. In the exercise, the human players are charged with trying to "rescue" an ineffectual but unstoppable GP system (which is the other "player"), which has been set up to only use "random guessing"—but they must do so by *incrementally modifying the search process without interrupting it*. The exercise is a game for two players, plus a Facilitator who acts as a referee. The human "User" player examines the state of the GP run in order to make amendments to its rules, using a very limited toolkit. The other "player" is the automated GP System itself, which adds to a growing population of solutions by applying the search operators and evaluation functions specified by the User player. The User's goal is to convince the System to produce "good enough" answers to a target supervised learning problem chosen by the Facilitator. To further complicate the task, the User must also provide the Facilitator with convincing justifications, or *warrants*, which explain each move she makes. The Facilitator chooses the initial search problem, provides training data, and most importantly is empowered to *disqualify* any of the User's moves if unconvinced by the accompanying warrants. As a result, the User is forced to work around our field's most insidious habit: that of "stopping it and starting over again with different parameters". In the process of working within these constraints, the participants—Facilitator and User—are made mindful of the habits they have already developed, tacitly or explicitly, for coping with "pathologies" and "symptoms" encountered in their more typical work with GP.

Keywords Mindful practice • Design process • Coding kata • Praxis • Mangle of Practice

W.A. Tozier (✉)
Ann Arbor, MI, USA
e-mail: bill@vagueinnovation.com

© Springer International Publishing Switzerland 2016
R. Riolo et al. (eds.), *Genetic Programming Theory and Practice XIII*,
Genetic and Evolutionary Computation, DOI 10.1007/978-3-319-34223-8_4

1   Why: An Excuse

More than a decade ago, Rick Riolo, Bill Worzel and I worked together on a genetic programming consulting project. As we chatted one day, one of them—I don't recall which, and accounts vary—was asked what he'd most like to see as GP "moved forward", and said he'd want the field to think more about the "symptoms" we so often see when we apply evolutionary search processes in complex settings. That is: premature convergence, slow and spotty improvement, a catastrophic lack of diversity, and more generally that ineffable feeling all GP professionals experience when we look at results and *know we have chosen unwisely.*

Now the reader will point out that the literature in our field overflows with well-written papers describing tips for avoiding local minima, improving on common search operators, and running "horse races" between Bad Old and Better New search methodologies applied to benchmark problems. But while as a rule these are presented as a sort of *general principle*, most in practice are *case studies* in which it is shown, for example, that search operator X acts under contingency Y and sometimes produces outcome Z.

I am not left with a sense that this catalog addresses my colleague's stated wish. While it is surely necessary to compile such a list of individual observations under particular suites of experimental treatments and benchmarks, it is insufficient until we can achieve the skill of recognizing *when* something noteworthy is happening. In particular, I want to focus on those frequent but contingent situations in which we are willing to say our GP system *resists our expectations.*

A good deal of this chapter will be spent explaining just what I mean by this particular usage of "resistance", but for the moment the idea will be clear enough from this simple mental exercise: What are things GP has "done"—in the context of a particular project—that have left you feeling unsatisfied, confused, or frustrated? Not because you've made a simple mistake setting a parameter, or introduced a bug in a codebase, but because (at that moment) you have no idea why the GP process is doing *that thing* under those particular circumstances? Maybe it's not converging when you know it should; maybe it's exploring unexpectedly complex algorithms rather than obvious simpler ones, despite your reasonable parameter settings; maybe it's failing to solve simple problems but easily solving hard ones. In any case, the system "resists" your careful *and knowledgeable* expectations by inexplicably refusing to follow your plan.

My general point, and the particular point of this chapter and the exercise it describes, is that we should then ask: *When we feel GP is resisting us, what should we do—and why?* Further, after more than 20 years of work in the field, I think we have learned enough for this to be reasonable and productive research program.

I'm sure every reader has at least one story they can tell, in which GP has "resisted" in this sense. I want to draw our attention to these incidents from a conviction that it is exactly the steps we take to *accommodate* GP's resistance which provoke our sudden insights into the structure of our problem, drive us to build a "workaround" that becomes a novel selection algorithm or architecture, and lead us to successfully re-frame our project to use a completely new approach. Almost any chapter in this series of GPTP Workshop Proceedings will contain a story just like this: We create an age-layered population because working without age-layering didn't perform as expected (Hornby 2009); we invent a new selection mechanism because traditional algorithms worked as described, but failed to capture crucial details of our problems (Spector 2012); we create an entirely new representation and suite of search operators because our goals aren't met by snipping up and recombining in the traditional way (Ryan and Nicolau 2003); and so on.

The point in each case is: one gains little *insight* into a problem when GP quickly pops out the "right answer" without a fight. Too often I see papers treating this ubiquitous resistance as something to be eradicated *before* "real users" are allowed run their own GP searches. Instead I'll argue here that "surprises" and "disappointments" are not only inevitable but *are the main source of value in many GP projects*, and as such should be the main focus of our theoretical and practical work.

Every interesting project will resist our plans and expectations. But in our field, whenever we're faced with even a bit of this sort of behavior, our first reaction is generally to act as though something has "gone wrong in the setup", then shut the misbehaving run down so we can start again with "better parameters" next time. I find it unusual for anybody to interrogate the system *in itself* when it resists, or to make an attempt to adapt or accommodate perceived resistance. I hasten to say this is not a fault with our field, but rather a symptom of broader philosophical and cultural problems in our approach to programming and computational research projects, and our understanding of computing more generally.

The exercise I describe in this chapter is intended to bring our attention back on GP as a dynamical process *in itself*, as opposed to a tool to be adjusted "in between" applications. It may be the case that other "traditional" machine learning methodologies are built on better-defined information-theoretic foundations, and come with suites of strong statistical tests for overfitting and robustness; as a result it's perfectly reasonable to treat them as *tools*, and as *malfunctioning* or just being *wrong for a problem* when they act in unexpected ways. But GP is somehow a different sort of animal: when it "resists", we are left with an essentially unlimited choice of how we should *accommodate* that resistance.

1.1   On Mindful Exercises

The habit of pursuing *kata*, "code retreats", "hackathons" and other skill-honing practices is popular among software developers, and especially among the more advanced. It apparently arose independently among a few groups in the 1980s, but Dave Thomas seems to have first used the Japanese term (Thomas 2013).

Indeed, the title of my exercise ("GP As If You Meant It") is directly inspired by from an exercise designed by Keith Braithwaite, "TDD as if you meant it" (Braithwaite 2011). One of the interesting aspects of Braithwaite's exercise is that it feels subjectively "harder" when attempted by advanced programmers honing their development form; he suggests that novice programmers haven't learned ingrained but questionable habits, and haven't identified "shortcuts" that "simplify" the practices.

In the same way, the exercise I describe will feel *most* artificial and restrictive to those of us with the most experience with GP. But like the martial arts exercises by which software *kata* first were inspired, it isn't intended to be simple or even pleasant for the participants. Just as Braithwaite's exercise targeted what he calls "Pseudo-TDD", I intend mine to surface habits we can think of as "pseudo-GP": the sense that it's cheap and painless to *just shut it off and start over* when we start sensing a problem is arising.

1.2   Caveats

What I describe here should be considered a "thought-experiment" backed up by my own experiences; there are only a few sketchy manual implementations to date. Some readers have imagined that the "automated" System player must somehow be a self-contained and tested process running on the cloud (and that is an eventual goal), but in all my early tests that role has been played by a handful of simple Ruby scripts, edited between "turns" by the Facilitator to reflect changes made by the User player. In other words, *there need not be an actual autonomous "computer player" in this game*, unless you feel compelled to write one up beforehand, and the game can as easily be played by a Facilitator who *codes* the moves of the System and *plays that role*, plus the User as described.

Further, there is no reason the User player need be a single human being. Indeed, it's common practice in many software development *katas* to work in teams or even as an aggregated crowd of everybody in the room. Many hands make light work, especially on a tight schedule. In other words, these roles are intended to be "notional" rather than definitional.

I should also point out that this game is not a serious suggestion for a new way of working on "real problems", nor as any sort of "training" for newcomers to the field of GP. Rather it is designed as a rigorous and formal exercise in mindful practice, to be undertaken by people already working closely enough with GP systems to recognize the problems when they arise.

2   "TDD As If You Meant It"

Keith Braithwaite seems to have first described this training exercise for software developers in 2009 (Braithwaite 2011). His target was a sense that software developers who thought they were using test-driven development practices (Beck 2002) were in fact doing something more like "Pseudo-TDD", a sort of slapdash and habitual approximation lacking many of the benefits of mindful practice.

While I've noted elsewhere that several agile software development practices share useful overlaps with the problems of GP science and engineering,[1] in this work I'll focus on those of TDD. In particular, the observation that test-driven development (or more accurately "test-driven design") *when done correctly* can break down the complex design space of a software project into a value-ordered set of incremental test cases, focus developers' attention on those cases alone, inhibit unnecessary "code bloat" and feature creep, and produce low-complexity understandable and maintainable software.

TDD *as such* is a very constraining and rigorous process—to the point where it can easily be described as "painful" (though also "useful") by experienced programmers. The steps are deceptively easy to gloss, misunderstand or miscommunicate, especially for those whose coding habits are ingrained. To paraphrase Beck and Braithwaite:

1. Add a little (failing) test which exercises the next behavior you want to build into your codebase
2. Run all tests, expecting *only the newest* to fail
3. Make the minimal change to your codebase that permits the new test to pass
4. Run all tests, expecting them *all* to succeed
5. Refactor the codebase to remove duplication

Even though a single cycle through this iterative process can take less than a minute, each step can throw itself up as a stumbling block for an experienced programmer. But the most salient for us here is the iterative flow of implementation (or "design") that the cycle imposes: it begins with a choice of *which little test should next be added*, and ends with a rigorous process of refactoring, not just of the new code but of the *entire cumulative codebase* produced so far. The middle three steps—implementing a *single* failing test and modifying the codebase *by just enough* so that all tests pass—feel when one is working as though they could be automated easily. The *mindfulness* of the process lives in the choice of next steps and (though somewhat less so) of standard refactoring operations.

Braithwaite's exercise does an interesting thing to surface the formal rigor of those decisions, by making them *harder* rather than easier. In "TDD as if you meant it", the participants (willing, of course) are asked to implement a nominally simple project like the game of Tic-Tac-Toe using TDD, and are given a list of requisite features and an extra constraint. Rather than using "normal" TDD and producing a suite of tests to exercise a separate and self-contained codebase, they are forced to add code *only* to the tests themselves (to make them pass), and can only produce a

---

[1] I imagine there is an Engineering Studies thesis in this for some aspiring graduate student: Genetic programming and agile development practices arose in the same period and more or less the same culture, and both informed by the same currents in complex systems and emergent approaches to problem-solving.

separate "codebase" when duplication or other "code smells" drive them to refactor the code already added to tests. In other words, the *demand for a warrant* for writing code is much more stringent.

Throughout the exercise, a facilitator patrols teams of participants and deletes *any and all code not called for by a pre-existing failing test.* Words like "irritating" and "annoying" crop up in participants' accounts of this onerous backtracking deletion the first few times it happens, as one might imagine. But as Adzic (2009) has said in descriptions of workshops he's facilitated, the resulting designs even for well-known algorithms in this artificially amplified setting become much more "open-ended" than would be expected if the code were written under the offhand attention of an experienced programmer without painful constraints.

Adzic passes along some observations in his account of a Tic-Tac-Toe exercise (Adzic 2009) that are especially interesting for me:

> By the end of the exercise, almost half the teams were coding towards something that was not a 3 × 3 char/int grid. We did not have the time to finish the whole thing, but some interesting solutions in making were:
>
> - a bag of fields that are literally taken by players—field objects start in the collection belonging to the game and move to collections belonging to players, which simply avoids edge cases such as taking an already taken field and makes checking for game end criteria very easy.
> - fields that have logic whether they are taken or not and by whom
> - game with a current state field that was recalculated as the actions were performed on it and methods that could set this externally to make it easy to test

In other words: innovative approaches to the problem at hand began to arise, though there wasn't enough time to finish them in the time allotted for the exercise.

3 GP As If You Meant It

In the same way that Braithwaite's coding exercise uses an onerous extra constraint[2] to drive participants towards more mindful and insightful decision-making, in this exercise I will demand a *warrant* for each implementation decision that moves a running GP setup away from random guessing. Braithwaite's target of "Pseudo-TDD" suggests an analogous "Pseudo-GP": one in which the fitness function and *post hoc* analysis is the only "interface" with the problem itself, and where the representation language, search operators, search objectives and other algorithmic "parameters" are *fixed* in the course of the run.[3]

---

[2]In this the sensibility reminds me of the constraint-driven art collective Oulipo (Becker 2012), who are perhaps most famous for the *lipogram*, a literary work which cannot use a particular letter of the alphabet.

[3]Braithwaite's participants (Braithwaite 2012) often acknowledge they *know* and *use* TDD as it's formally described, but rarely take the time to do so unless "something goes wrong". I imagine many GP users will say they *know* and *use* all the innumerable design and setup options of GP, but

Not only will traditional search operators like crossover, mutation and [negative] selection not come "for free" in this variant, but in every case we must develop a cogent argument in favor of starting them as part of an ongoing search process. Similarly, the initial selection criteria will be limited to a single training case, and expansion of the active training set will have to be made *in response to particular features* of observed progress, not merely on the basis of the assumption that "more will be better".

The result is a painfully incrementalized process, one that focuses on the refinement and eventual correction of an unstoppable search which was *intentionally* "started wrong", and which must be carried out by *doing surgery on the living patient* to correct perceived "pathologies" and "resistance". Along the way, a fraction of the mysteries of "pathology" has the potential to be much clearer and better-defined.

4   Overview

The exercise is structured as though it were a game for two players, plus a Facilitator who establishes the ground rules, provides any needed technical infrastructure, and acts as referee. One player is (or represents) a running GP System, and one player is (or represents) a human User trying to mindfully drive the System's performance in a desirable direction over a series of turns.

While I speak below of the System player "being" a self-contained software process, it is of course a loose role that might more easily be played by another human, potentially the Facilitator herself, writing and running a simple series of scripts on a laptop. Similarly, while I may say that the User player "is" a single human being, it could as easily be a room-full of students or workshop participants, or a mailing list voting over many weeks on strategies for each turn. Indeed, in working out the exercise as it's described here, I've "played" all the roles myself, simultaneously, and still found interesting and unexpected insight.

In preparing the *kata*, the Facilitator selects a *target problem*, which should be a supervised learning task for which plenty of data is available. The target should not be "toy" in the sense of having a simple, well-known answer; rather it should be challenging and open-ended enough to warrant a publication if solved (a problem that has *recently* been solved might do in a pinch, though there is no shortage of open ones). Any good book of mathematical recreations [for example (Winkler 2003) or nearly any book by Martin Gardner, Ian Stewart or Ivan Moscovich] will provide numerous abstract problems that have never been attempted with GP.

---

treat them as adjustments to be invoked only when "something goes wrong". I offer no particular justification for either anecdote here, but the curious reader is encouraged to poll a sample of participants at any conference (agile or GP).

The Facilitator will also need to choose a representation language, set up an initial Tableau, and provide enough software infrastructure so that the System's turns can be made easily. In each turn of the game, both the System and the User players take actions to modify and extend the initial Tableau: the User's options include adding *rubrics* and *operators* (described below), and in its turns the System acts "mindlessly" by *invoking* the specified rubrics and operators to add a fixed number of new individuals to the growing population.

The User's goal is to drive the System towards producing *sufficiently good solutions* to the target problem, but their decisions are constrained by the obligation to provide a *warrant* for every change made which is convincing to the Facilitator. The particular definition of a "sufficiently good solution" is left for the players and Facilitator to decide in context: it may be impossible to completely "solve" some target problems, given the tools at hand.

It should always be kept in mind that *the point of the exercise is to evoke interesting and useful warrants*, not merely to drive the System in a desired direction. In each turn the User is obliged to produce a convincing *warrant* for every move she makes, which must be reviewed and approved by the Facilitator before play proceeds. These warrants need not be *factually correct*, but they must be *convincing* in the context of the game state at the time they're put forth.

4.1 The Tableau

The game "board" is a *Tableau* with two components: A list of search `operators`, and a two-dimensional spreadsheet-like table which uses `answers` as its row labels and `rubrics` as its column labels. Initially the Tableau is empty, except for a single entry on the list of `operators` labeled "random guess".

The "random guess" operator is set up by the Facilitator before play begins; it produces a single random `answer` for the target problem, with no arguments. Think of it as equivalent to the "initialization" function used to produce a single "random" individual in a traditional GP setting.

During the User's turn, she can examine the state of the Tableau, including any or all of its history, and the algorithms in play, and apply any amount of data analysis or statistical work she wants. On her turn she is permitted no more than *two* moves: She can (optionally) code and append one new `operator` to the list of `operators`, and she can (optionally) code and append one new `rubric` column to that table.

During the System's turn, it will produce a fixed number of new `answers`. It creates each new `answer` by first picking an entry from the list of `operators` with uniform probability. When an `operator` is chosen, it executes the indicated algorithm, selecting parents from the `answers` table as needed. For each required selection of a parent, the scores recorded in the table of `answers` vs `rubrics` are used (see below). Once parents are selected, the `operator` is applied and a new `answer` is immediately appended to that part of the Tableau.

4.1.1 Answers

What I'm calling an `answer` here would probably be called an "individual" in the GP literature. In this case, it is a particular *script* or program in the representation language the Facilitator has chosen for the target problem.

Answers never "die", and cannot be removed from the Tableau by either player.

4.1.2 Operators

Each `operator` is a function which takes as its argument an unordered collection of zero or more `answers`, and which produces a new collection of one or more `answers` as output.

The initial Tableau includes only a single `operator`, which implements a pre-coded "random guessing" algorithm. On the User's move, she may build and launch other more complex (and familiar) `operators` like crossover or mutation. The only permitted argument is a set of zero or more `answers`; no numerical or other parameters are allowed. A wide variety of `operator` algorithms are still possible, and the details of the code in which they are implemented is left to the Facilitator to specify as part of the game setup.

Except for some unusual edge cases, the *specific* set of `answers` to which an `operator` will be applied cannot be chosen directly by the User. All "parents" defined for each `operator` are chosen independently (and incrementally) by *lexicase selection*, using the suite of `rubrics` in play when the System takes its turn. This lexicase selection process samples every `rubric` in the Tableau with equal probability. It may be possible for the User to design a new "selection" `operator` which takes as its argument "all the answers" and somehow culls that collection down to a subset—but to do so, it must rely on the immediate state of the `answers` it is given; `rubric` scores are "stored" only in the Tableau answer table itself, not as gettable attributes of the `answers`, making it very difficult for any new `operator` algorithm to use `rubric` scores in its implementation.

Note again: no mechanism exists which *removes* answers from the Tableau.

4.1.3 Rubrics

A `rubric` is a function which returns a scalar value (not necessarily a *number*) for any given `answer`, conditioned (as needed) on the instantaneous Tableau state. When a `rubric` is applied to an `answer`, it sets a new value (or "score") for that `answer` in the appropriate cell of the "spreadsheet" portion of the Tableau.

No score is ever changed for an `answer`, once it has been set by a `rubric`. However, if multiple `answers` exist which have the same `script`, each may be scored at different times or in different contexts, and the resulting values may differ.

An aggregate or *higher-order* `rubric` can be created, but its existence "entails" all of the component `rubrics` from which its score is derived. So, for example, if the User constructs a rubric like "maximum absolute error observed over any of these 30 training cases", on the turn when she adds that rubric up to 31 new columns will be added to the spreadsheet portion of the Tableau: one for each of the 30 requisite "sub-rubrics", which produce as their scores the absolute error for a single training case, and also one *aggregate* rubric which calculates and reports "maximum of those other 30 scores". Whenever the User submits such a higher-order rubric on her turn, all of the entailed rubrics are added automatically if they do not already exist. Note however that there must still be a *warrant* for the rubric, which convinces the Facilitator of its potential usefulness.

The User can use any information present in the Tableau in building new rubric functions (such as row numbers, string values of answers, or scores), and can also examine the detailed state of the interpreter before and after the script is run. It should be clear therefore that the User can specify rubric functions which score aspects of the problem such as:

- the absolute error for a single training case
- the number of tokens in the answer's script
- maximum error measured in any of 35 other rubrics
- number of div0 errors produced when running a script with a particular set of inputs
- number of stochastic instructions appearing in the answer's script
- rewrite difference between the scored script the most common answer in the population
- (and so on)

Various problem-specific aspects are glossed here, and it is left to the Facilitator to be reasonable in the context of the target problem and the representation she has chosen. Suffice to say, the construction of useful rubrics in response the System's moves is the core of the User player's game strategy.

4.2 Lexicase Selection

All operator inputs are chosen by lexicase selection. Slightly simplifying Spector's original description (Spector 2012), the following algorithm can be used:

- (beginning with a "set under consideration" which includes all answers...)
- for each rubric in a random permutation of all the rubrics in the Tableau, discard all answers from the set under consideration whose score on this rubric is sub-optimal (relative to the current set under consideration)
- if multiple answers remain after all rubrics have been applied, return one answer picked randomly, with uniform probability, from the remaining set under consideration

At the beginning of the game when no rubrics have been added and no operators exist which require "parents", no selection occurs or needs to occur.

But note that the algorithm as described would *still* provide a parent, even when no `rubrics` have been specified: an empty set of `rubrics` would be immediately exhausted without eliminating any `answers`, and then a single `answer` would be selected with uniform probability from that complete set.

Lexicase selection has several characteristics which argue for its use here; in particular it is interesting because it seems (anecdotally, but see also Helmuth in this volume) to be a relatively "slow" selection method, permitting a diversity of `answers` to coexist in a population at the same time. However I admit readily that I've specified it as a core part of this *kata* because it is *unfamiliar to most prospective players*, and therefore more likely to produce unexpected behavior of a useful sort.

4.3   The User's Turn

In her turn, the User can do either (or both) of the following:

1. add one `operator` to the list in the Tableau
2. add one `rubric` to the Tableau, which can be "higher-order" and therefore entail others

To support these decisions, the User can examine the Tableau state and history in detail. Before any code is implemented, though, a *warrant* must be written for each one.

4.3.1   Warrants

A *warrant* is a verbal or written argument which spells out the justification for one of the User's moves *in the context of the game as it stands when the move is made*.

Suppose, for example, that the User has not yet added *any* `rubrics`, and would like to add a new one that scores `answers` for a single training case. Since the System is simply making new `answers` with the "random guess" `operator`, it seems perfectly reasonable to warrant this new `rubric` simply by pointing out that selection can't drive search towards better answers without at least one training case being used.

Suppose further that later in the same game the User has added more `rubrics`, such that more than 100 training cases are being used to select parents for the several `operators` in play. In this context, the reasons given for adding a new `rubric` that specifies *just that one particular training case* must surely be very different: perhaps there is a problematic region of the response surface, or a different training case that needs additional "support" to differentiate between two close input states (and so forth).

That said, a warrant does *not* need to be "technical" or even "rigorous", but merely robust enough to be convincing *in the moment*. For example, "I made a crossover `operator` because I think we need to search for new answers 'in between' the parents," sounds to me like a shoddy excuse that invokes received wisdom. On the other hand, "I made crossover so we can drop the variance on this rubric and foster inbreeding of *these* solutions *here*," given a glance at the Tableau and charts on hand, should be more convincing.

Note though that decisions and the warrants that supported them may well prove to have been wrong in hindsight. This in itself is an interesting and useful outcome in the game: when an earlier decision does not actually produce the expected effect in the system, this is a perfect point for the User to be reminded of the apparent inconsistency. Such "failed" warrants shouldn't be rescinded, but it may be appropriate to bring them up as "concerns" in later moves, especially if they begin to accumulate.

There is one particular warrant that the Facilitator should always permit, as long as it isn't abused by the User: "Because I have no idea what will happen when I do this, but I suspect it may be useful in order to [X] later on." In other words, it is perfectly reasonable for the User to state outright they are "exploring" the range of system behaviors. More generally, the basis on which the User, and therefore the Facilitator, decides to make or permit a change should always depend on the history of the game so far. Indeed, reasoning will *always* change dramatically over the course of any interesting game, under the constraint that the Facilitator should never approve a change on the argument that "that's the way we always do it". Every clear and convincing *why* argument will surely be contingent on the immediate state of the system and dependent on all the prior decisions made by the User and Facilitator.

4.4   The Facilitator

The Facilitator is responsible for picking a representation language and target problem, for providing technical infrastructure as required for the *kata*, and for approving and implementing any changes to the System that the User player makes. As noted several times above, *warrants* are the focus of the exercise, and as such great care should be taken that each decision is well-justified.

4.4.1   Choosing a Representation Language

No particular constraints apply to the language or representation chosen for the exercise, except that it should be sophisticated enough to support *redundant capacity*. That is, for any given algorithmic goal, there should be multiple paths to success. So for example if the problem's solution could reasonably be expected to involve ordered lists, then it would be best if the language had at least two different ways to "use lists", for example with iterators, recursion, a comprehensive set of second-order functional operators, an explicit List type with associated methods, and so forth.

If the language is sophisticated enough to have "libraries" of instructions and types intended for specialized domains, I'd strongly encourage that *all of these should be used*. That is, the Facilitator should err on the side of "winnowing complexity", rather than forcing the User (and thus the System) to *invent basic data structures* or the idea of floating-point numbers at the same time they're trying to solve the "real" problem under consideration.

Finally, if there is a choice between a familiar language and an odd one, then the odd one should be chosen, all other things being equal. Unfamiliarity can be a useful constraint here.

4.4.2 Target Problems

As with the choice of language, in the choice of problems the planner should aim for something that would strike any experienced GP person as "ambitious". Which is to say: a reasonably good programmer would be able to hand-code the answer with a day's thought and work, but only in a familiar language; in an oddball GP language, it should feel "practically impossible" to hand-code. That said, it should *also* be clear to any programmer familiar with the language specification that the needed *components* are all there, and that there are enough "parts" to approach any sub-task that crops up along the way from more than one angle.

Colleagues have pointed out that even if the chosen problem turns out to be "too simple", in the sense that the System solves it quickly without much input from the User, then the "problem" addressed in the exercise can be extended by the Facilitator to one of driving the System to find a second *dissimilar* solution... still without restarting the search process, of course. Keeping in mind that the purpose of the exercise is to provoke insight into the justification for particular decisions, this decision (by the Facilitator) can be justified by being a perfectly reasonable event in a real research project, in which intellectual "stretch goals" are commonplace when resources permit.

4.4.3 Initial Setup and Restrictions

- the only `operator` is "random guess", which creates one new `answer` with an arbitrary script
- no `rubrics` are present
- the System player moves first
- there is no mechanism for *removing* `answers` from the tableau
- the System player *always* uses lexicase selection, *always* chooses `operators` with equal probability from the current list, and *always* uses all `rubrics` in the Tableau
- a `rubric` can only be *run* on a given `answer` once; stochastic scripts will only be sampled one time, and no `rubric` score is ever recalculated after the first time, though multiple copies of the same stochastic `answer` will probably end up with different scores in the same field.

4.4.4 The System's Turn

During its turn, the System player will add a specified number of new `answers` to the Tableau, one at a time, using a simple form of lexicase selection. Until it reaches its halting state it iterates this cycle:

1. select one `operator` from those in the Tableau, with equal probability
2. apply lexicase selection to select the required number of input `answers`
3. apply the chosen `operator` to the inputs `answers` to produce one or more new `answers`, and append those new `answers` immediately to the Tableau
4. `HALT` if the number of new `answers` meets or exceeds the limit, and delete any extra `answers` that exceed the limit; otherwise, go to step (1).

The number of `answers` created in each turn should be enough to have a chance of providing new and useful information to the User, and not so much that the System state grows out of control. I would suggest 100 or 500 `answers` per turn; this is about the size of the typical "population" for most GP users, and is on a comfortable scale for them.

5 Why: A Warrant

Genetic Programming[4] embodies a very particular *stance* towards the scientific and engineering work of modeling, design, analysis and optimization. I increasingly suspect that social resistance to GP has little to do with the quality of our technical results. Rather it arises from unfamiliarity with GP's very particular "way of working". We in the field have become used to it—perhaps to the point of taking it for granted—but colleagues in other fields have not.

Briefly, the systemic fault lies in the awful "scientific method" that permeates our cultural dialog about the practice of science and engineering. You know the one, which I can here as something like:

vision → planning → design → architecture → implementation → testing → debugging

I'm sure very few scientists or engineers of my acquaintance would admit any *real* project has *ever* followed this narrative in a literal sense. But that story nonetheless informs and constrains much of our work lives, from fund-raising to publishing reports, producing narratives of our work that run more or less like this: "Based on the body of published work, an insight was had. The insight was framed as a formal hypothesis. The hypothesis (shaped by current Best Statistical Practices) immediately suggested an experimental design, which design is obvious

---

[4]And not just Genetic Programming as such, but also the broader discipline to which I claim it belongs and which is not obliged to be either "genetic" or "programming". I prefer to call this looser collection of practices "generative processing", and will also abbreviate it "GP"; assume I mean the latter in every case.

to anyone familiar with Our Discipline. That experimental design was undertaken, the data were collected, the hypothesis duly tested, and now we can be confident of its veracity because... well, you just heard me say 'Best Practices', right?"

"Nothing surprising happened while we were working on this project," in other words. Under trivial term substitutions—"cost–benefit analysis" and "requirements document" for "hypotheses" and "experimental design", for example—the same narrative can be used to describe almost any institutional project management or public policy planning process as well. The flow in every case is essentially from *vision* to *plan*, *plan* to *implementation*, *implementation* to *verification*, and *verification* to *validation*.

Of course, nobody "really believes" this narrative who has ever done the work. It is a matter for another day to draw parallels with the social construction of religious belief.[5] And I am not the first to point it out; the history of Philosophy of Science is built primarily from the numerous philosophical challenges to this artificial narrative, from Peirce and Dewey nearly a century ago, to Kuhn and Lakatos and Feyerabend in the 1970s, and with many more to be found in the Table of Contents in any Philosophy of Science text.

That said, it is Andrew Pickering who has provided my immediate inspiration for this project.

5.1 On the Mangle of Practice

Andrew Pickering's monograph *The Mangle of Practice* (Pickering 1995) is a decade old, but surprisingly little-known outside his discipline of Science Studies. His approach is especially useful here, because I find it captures a surprising amount of our *actual experience* of building and using GP systems. Indeed, most colleagues who hear it for the first time utter an inevitable "didn't we already know this?"

Pickering's approach focuses on that problematic division I've sketched above, between the illusory (but publishable) linear narrative of the "scientific method", and the realized experience we all have had of *performing science* (or Mathematics, or Engineering, or for that matter Art). At the cost of glossing too much of his well-considered structure, let me summarize.

First, I should remind us all that the *performance of science* is just that: not an isolated but perceptive mind standing apart from the world, working in an objective and static field of "externalities" and "facts", but a *performance* done by a human being present in that world. In Pickering's framework, we can say that research proper begins only when the researcher makes some artifact or formal "machine" in the world: writes a block of code, designs a technical instrument, considers a particular equation, draws a pencil sketch, or simply has a thoughtful conversation

---

[5]Paul Veyne's excellent *Did the Greeks Believe in Their Myths?* (Veyne 1988) might be an interesting starting point, I suspect.

at a conference. Let me call this artifact *the thing made*. This is not the scientific paper that results the end of the project, but rather the sum of all the sketchy notes, the cloud of more-or-less coherent ideas, the code and instruments, the collected observations, the plan and the community of colleagues helping with that plan: everything done in the world, mentally or physically, towards the goals of the project.

Pickering's model jumps quickly away from more traditional "scientific methods" when he treats this mechanism as capable of *agency in its own right*, and is willing to say that it can and does *resist* our intentions. In the context of the Mangle, the *thing made* is the conduit of the facts of the actual world to the researcher (and also of the cultural assumptions and norms of one's discipline, of the inherent tendencies of the raw materials and the practitioner's toolkit). "Resistance" here is not merely a reference to a software bug, a mathematical mistake or a shortage of crucial raw materials, but specifically denote one's sense *on seeing it* that "something's not quite right". In other words, it is the *thing made*'s resistance "on behalf of" the real world which forces the researcher to reconsider, change or adapt her plans, or otherwise *accommodate* that resistance.

The inspiration for granting agency to machinic abstractions (or even concrete dynamics) is obvious whenever we hear the phrases we utter in the course of our work: the system "is acting up"; the mathematics is "pointing something out"; the machine "wants to do X instead of Y". Projects in science, engineering and the arts do not proceed from a stage of planning to a stage of implementation, except in the ahistorical mythology of our published papers (see Koutalos 2008 for a particularly good assessment of this from a biologist). Pickering's Mangle[6] does much better at capturing our first-hand experience of the work as an emergent dance of human and machinic agency *with one another*. The researcher starts to follow her vision by making (and altering) some artificial thing, that *thing made* acts as a channel for the world itself to resist, and as a result the researcher *accommodates* that resistance by moving in some different direction. In the traditional linear narrative, we elide the work as it and re-frame it as a sort of idealized, apersonal Platonic truth: we use the passive voice, we hide the missteps and confusion, after the fact paint a story which flows from vision to plan to success. But within the dance of Pickering's Mangle, the degree to which we as researchers can successfully *accommodate* the resistance

---

[6]Pickering's word "mangle" and the way he came to choose it are a recursive example of the framework itself:

> ...I find "mangle" a convenient and suggestive shorthand for the dialectic because, for me, it conjures up the image of the unpredictable transformations worked upon whatever gets fed into the old-fashioned device of the same name used to squeeze the water out of the washing. It draws attention to the emergently intertwined delineation and reconfiguration of machinic captures and human intentions, practices, and so on. The word "mangle" can also be used appropriately in other ways, for instance as a verb. Thus I say that the contours of material and social agency are mangled in practice, meaning emergently transformed and delineated in the dialectic of resistance and accommodation....

we encounter is *exactly* the degree to which we can say we have made progress in our projects.

In a GP setting, the notion of "machinic agency" seems much closer to our experience; after all, we are obliged not only pick or write a specialized formal language to represent the space of solutions, but in every project we must also cobble together some framework of search operators, fitness operators, algorithms and instrumentation. But even when we've written all the code and set all the parameters personally—dotted the Ts and crossed the Is, as it were—we're *still* driven to speak of our GP run "doing" things, rather than merely unfolding according to our plan. Indeed, if it happens by chance that GP "runs according to our plan" then arguably the problem was too *boring* to be worth mentioning. . . .

I will argue below that GP's power (and difference from other machine learning approaches) lies in the very particular form of resistance it can offer us as its users. This is not merely "resistance" of the frustrating kind: we use GP most effectively when we want it to surprise us. The "surprise" is certainly something we are forced to accommodate with just as much attention and concentration as any more annoying resistance which might be thrown up, for example when we are forced to figure out how the "winning solution" GP has disgorged actually works.

It is worth saying explicitly now (and then again as many times as necessary) that by granting the *thing made* a machinic agency of its own, we can frame the problem of "pathology" and "symptoms" in GP more constructively. A GP system does not resist by "having the wrong population size" or by "having too high a mutation rate"; those are not *behaviors*, but tiny facets of a complex plan instantiated (to some extent) in a complex dynamical system. Rather we should say that a GP system is resisting when it is *raising concern or causing dissatisfaction in its human user*. It is inevitably that *human* observer who is driven to the insight needed to provide an accommodating response.

5.2   GP as "mangle-ish Practice"

The broader field of machine learning seems to take a much more "linear" stance towards its subject matter than we do in ours, in the sense that the pastiche of the "scientific method" applies. The result of training a neural network or even a random forest on a given data set is not expected to be a *surprise* in any real sense, but rather the reliable and robust end-product of applying numerical optimization to a well-specified mathematical programming problem. Indeed, the supposed strength of most machine learning approaches is the very *unsurprising* nature of their use cases and outputs.

On the other hand, we all know that GP embodies a capacity to *tell us stories*, even in the relatively "simple" domain of symbolic regression. The space under consideration by GP is not some vector of numerical constants or a binary mask over a suite of input variables, but the *power-set* of inputs, functions over inputs, and higher-order functions over those. We who work in the field can be glib about the "open-endedness" of GP systems, but that open-endedness puts GP in a qualitatively different realm from its machine learning cousins. While GP can be used to explore arbitrarily close to some parametric model, its more common use case is exactly the production of *unexpected* insights.

When the GP approach "works", it does so by offering *helpful resistance* in our engagement with the problem at hand, whether in the form of surprising answers, validation of our suspicions, or simply as a set of legible suggestions of ways to make subsequent moves. GP *dances* with us, while most other machine learning methods are exactly the "mere tools" they have been designed to be.

5.3 Against Replication

Nonetheless, there seems to be a widespread desire inside and outside our field to frame GP as a way of exploring *unsurprising* models from data. As with neural networks or decision trees, the machine learning tool-user is expected to proceed something like this:

1. frame your problem in the correct formal language
2. "get" a GP system
3. run GP "on your data"
4. (unexpected things happen here, but it's not our problem)
5. you have solved your problem

This is of course exactly the stance expected in any planning or public policy setting, or any workplace using waterfall project management. And as we know from those cultures, "being surprising" could be the worst imaginable outcome. Given that pressure, it's no wonder that so much of GP research is focused on the discovery of constraining tweaks aimed at bringing GP "into line" with more predictable machine learning tools. If only GP could be "tamed" or made "adaptive" so that step (4) above *never happens.* ... I imagine this is why so many GP research projects strive for rigor in the form of counting replicates which "find a solution": they aim not to convince *users*, but rather to demonstrate to critical peers that GP can be "tamed" into another mere tool.

Think about "replicates" for a moment. What might a "replicate" be for a user who wants to exploit GP's strength of discovering new solutions? If one is searching for noteworthy answers—which is to say *surprising* and *interesting* answers—then a "replicate" must be some sort of proxy for user frustration in step (4) above. That is, a "replicate" stands in for a project in which search begins, stalls, and where the user cannot see a way to accommodate the resistance in context... and just gives up trying.

I cannot help but be reminded of the fallacy, surprisingly common both in and outside of our field, that "artificial intelligence" must somehow be a self-contained and non-interactive process. That is, that an "AI candidate" loses authenticity as soon as it's "tweaked" or "adjusted" in the course of operation. It is as if every new-born "AI" must be quickly jammed into an air-tight computational container and isolated *until it learns to reason by itself*—and for that matter without exceeding a finite computational budget.

If humans creating real intelligences (for example, other human beings) treated them anything like the way computer scientists insist we treat nascent artificial intelligences, I have no doubt that the resulting murder convictions would be swift and merciless. It is my hope with this contribution to suggest that we might be able to do better than the virtualized serial murder that is our legacy to date.

Consider the poor "GP user" that most of our research seems aimed at, one who is carefully "not interfering" with her running GP system: she can only peer at a results file after the fact, and can't fiddle with the "settings" while the thing is actually working. But of course during any given run of 100 generations, *all sorts* of dynamics have happened: crossover, mutation, selection, all the many random choices.

Imagine for a moment if she were given perfect access to the entire dynamical pedigree of the unsatisfying results she receives at the end, and were able to backtrack to any point in the run and *change a single decision*. Before that point, it's unclear how badly things will actually turn out at the 100-generation mark; at some point after that juncture, it's obvious to anybody watching that the whole thing's a mess. If such miraculous insights were available, then surely her strategy would be one of *intervention*: even if she could only decide after the fact, she should roll back the system to that crucial turning point, make a change aimed at avoiding the mess, and then continue from there.

Lacking (as we do) this miraculous insight, or the tools for understanding the internal dynamics of any particular GP system, on what grounds does it seem reasonable to stop *any* run arbitrarily at a pre-ordained time point and begin again from scratch? Replication, in the sense we are prohibited from reaching in and affecting outcomes, is no better than dice-rolling.

I would much rather say this: Insofar as GP *surprises us*, and since that is its sole strength over more familiar and manageable machine learning frameworks, we must learn to recognize and accommodate the surprises that arise in its use. Some surprises will always remain disappointments, but the senseless restriction we impose on engaging the systems we build blocks us from seeing others as **encouraging opportunities to improve our plans before it's too late**.

References

Adzic G (2009) Tdd as if you meant it – revisited. http://gojko.net/2009/08/02/tdd-as-if-you-meant-it-revisited/

Beck K (2002) Test driven development: by example. Addison-Wesley, New York

Becker DL (2012) Many subtle channels: in praise of potential literature. Harvard University Press, Cambridge, MA

Braithwaite K (2011) Tdd as if you meant it. http://cumulative-hypotheses.org/2011/08/30/tdd-as-if-you-meant-it/

Braithwaite K (2012) Tdd as if you meant it (workshop recording). http://www.infoq.com/presentations/TDD-as-if-You-Meant-It

Hornby GS (2009) A steady-state version of the age-layered population structure EA. In: Riolo RL, O'Reilly UM, McConaghy T (eds) Genetic programming theory and practice VII. Genetic and evolutionary computation, chap 6. Springer, Ann Arbor, pp 87–102. doi:10.1007/978-1-4419-1626-6_6

Koutalos Y (2008) The docile body of the scientist. In: Andrew P, Keith G (ed) The mangle in practice: science, society, and becoming. Duke University Press, Durham. https://www.dukeupress.edu/the-mangle-in-practice

Pickering A (1995) The mangle of practice: time, agency, and science. University of Chicago Press, Chicago, IL

Ryan C, Nicolau M (2003) Doing genetic algorithms the genetic programming way. In: Riolo RL, Worzel B (eds) Genetic programming theory and practice, chap 12. Kluwer, Dordrecht, pp 189–204. doi:10.1007/978-1-4419-8983-3_12. http://www.springer.com/computer/ai/book/978-1-4020-7581-0

Spector L (2012) Assessment of problem modality by differential performance of lexicase selection in genetic programming: a preliminary report. In: McClymont K, Keedwell E (eds) 1st workshop on understanding problems (GECCO-UP). ACM, Philadelphia, PA, pp 401–408. doi:10.1145/2330784.2330846. http://hampshire.edu/lspector/pubs/wk09p4-spector.pdf

Thomas D (2013) Codekata: how it started. http://codekata.com/kata/codekata-how-it-started/

Veyne P (1988) Did the Greeks believe in their myths?: an essay on the constitutive imagination. University of Chicago Press, Chicago, IL

Winkler P (2003) Mathematical puzzles: a connoisseur's collection. A K Peters/CRC Press, Natick, MA/Boca Raton

nPool: Massively Distributed Simultaneous Evolution and Cross-Validation in EC-Star

Babak Hodjat and Hormoz Shahrzad

Abstract We introduce a cross-validation algorithm called nPool that can be applied in a distributed fashion. Unlike classic k-fold cross-validation, the data segments are mutually exclusive, and training takes place only on one segment. This system is well suited to run in concert with the EC-Star distributed Evolutionary system, cross-validating solution candidates during a run. The system is tested with different numbers of validation segments using a real-world problem of classifying ICU blood-pressure time series.

Keywords Evolutionary computation • Distributed processing • Machine learning • Cross-validation

1 Introduction

The Age-Varying fitness approach is suitable for data problems in which evolved solutions need to be applied to many fitness samples in order to measure a candidate's fitness (see Hodjat and Shahrzad 2013). This is an elitist approach: best candidates of each generation are retained to be run on more fitness cases to improve our confidence in the candidate's fitness. The number of fitness evaluations in this method depends on the relative fitness of a candidate solution compared to others at any given point.

EC-Star (see O'Reilly et al. 2013) is a massively distributed evolutionary platform that uses age-varying fitness as the basis for distribution, thus allowing for easier distribution of large data problems through sampling or hashing/feature-reduction techniques, breaking the data stash into smaller chunks, each contributing to the overall evaluation of the candidates.

In this system, age is defined as the number of fitness samples a candidate has been evaluated upon. EC-Star uses a hub and spoke architecture for distribution, where the main evolutionary process is moved to the processing nodes (see Fig. 1). Each node, or Evolution Engine, has its own pool and independently runs through

B. Hodjat (✉) • H. Shahrzad
Sentient Technologies, 1 California St. #2300, San Francisco, CA, USA
e-mail: babak@sentient.ai; hormoz@sentient.ai

© Springer International Publishing Switzerland 2016
R. Riolo et al. (eds.), *Genetic Programming Theory and Practice XIII*,
Genetic and Evolutionary Computation, DOI 10.1007/978-3-319-34223-8_5

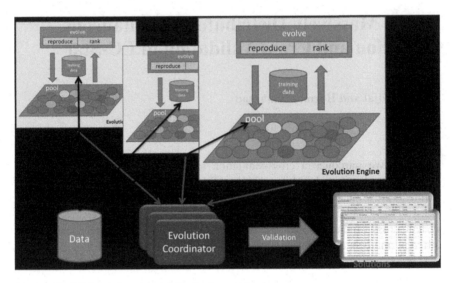

Fig. 1 The EC-Star hub and spoke distribution architecture

the evolutionary cycle. At each new generation, an Evolution Engine submits its fittest candidates to the server for consideration. This is typically after a set number of evaluations (i.e., the maturity age). Age is defined as the number of samples a candidate is evaluated upon.

The server side, or Evolution Coordinator, maintains a list of the best of the best candidates so far. EC-Star achieves scale through making copies of candidates at the server, sending them to Evolution Engines for aging, and merging aged results submitted after aging simultaneously on the Evolution Engines (see Fig. 2). This also allows the spreading of the fitter genetic material.

EC-Star is massively distributable by running each Evolution Engine on a processing node (e.g., CPU) with limited bandwidth and occasional availability (see Hodjat et al. 2014). Typical runs utilize hundreds of thousands of processing units spanning across thousands of geographically dispersed sites.

In the Evolution Coordinator, only candidates of the same age-range are compared with one another. This technique is called age-layering, and it was first introduced by Hornby (2006)—note, however, that the definition of age here is quite different. In EC-Star, each age-range has a fixed quota, and a 'shadow' of a candidate that has aged out of an age-layer is retained as a place-holder for filtering incoming candidates. To balance the load, a farm of Evolution Coordinators are used, all of which are synchronized over a single age-layered pool of candidates.

Typically, candidates harvested from the top age-layer of an EC-Star run are validated on an unseen set, post harvest, in order to ensure generalization. What if some validation could take place at scale in a distributed manner?

The nPool approach described in this paper is inspired by the well-known k-fold cross validation technique (see Refaeilzadeh et al. 2009), in which k iterations of training and validation are performed on k equally sized segments (or folds) of data,

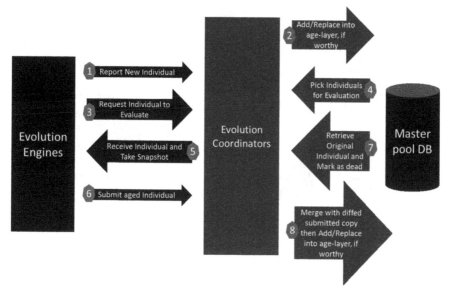

Fig. 2 Evolution Coordinators make copies of candidates and send them to the Evolution engines for further aging. The reported results are merged. Note that, in step 8, if the candidate is not worthy of being added to the new layer, it is discarded, but its 'dead' copy may remain in lower layers to help with filtering (step 7)

such that within each iteration a different fold is held-out for validation while the remaining $k - 1$ folds are used for learning. The learned models are asked to make predictions about the data in their respective (unseen) validation folds.

2 Description

Similar to k-fold cross validation, in nPool, the training data sample set is divided (roughly) equally into n segments. This division can be done randomly and segments should be mutually exclusive. Each evolution engine validates candidates exclusively on data samples from the segment randomly assigned to them by the Evolution Coordinator at start up. Every candidate generated by an Evolution Engine is tagged with the segment assigned to that Evolution Engine. We call this the candidate's originating segment, and at birth, the candidate's current segment is said to be its originating segment. Once a candidate has been evaluated on a sufficient number of data samples from its own segment, it is said to have hit its graduation age. The Evolution Coordinator tags graduated candidates with a segment id (out of the n available) that they have not been tagged yet. This id is the new current segment for the candidate.

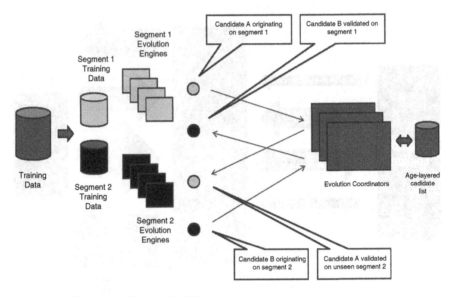

Fig. 3 Distributed cross*Tesladog*-validation

Evolution Engines ensure that candidates being sent down for further validation are only sent to Evolution Engines with a segment id equal to the current segment of the candidate. A candidate is said to have completed its validation once it has aged sufficiently on data samples from all available segments.

Evolution Engines validating candidates on segments other than the candidate's originating segment are barred from bearing offspring. This way, new generations are not contaminated by data from other segments, which we aim to keep as unseen for them for cross-validation purposes (see Fig. 3).

Rather than training on $k-1$ and validating on one segment, in nPool, we train on one and validate on $k-1$. This is in order to maintain complete exclusivity between the training and validation sets. Also, this ensures a more reliable assessment for the generalization of the candidates by using larger unseen validation sets. It is important, however, to ensure that the size of each segment is large enough to avoid over-fitting.

The segments should be mutually exclusive. In the experiments for this paper, the segments were divided up randomly. However, depending on the application, the division of the data into segments might require stratification.

3   Rule-Based Representation and a Real-World Problem

We demonstrate our approach on a real world problem of classifying time series of arterial blood pressure data. Our particular area of investigation is acute hypotensive episodes.

A large number of patient records are time series based. Some are at the granularity of high resolution physiological waveforms recorded in the ICU or via the remote monitoring systems. Given a time-series of training exemplars each of length T (in samples), to build a discriminative model capable of predicting an event, features are extracted by splitting the time series into non-overlapping, divisions of size k samples each, up to a certain point $h < T$ such that there are $m = h/k$ divisions. A number of aggregating functions are then applied to each of these divisions (a.k.a windows) to give features for the problem.

We use a decision list (Rivest 1987) representation as the model for the candidate in EC-Star. In this representation, each rule is a variable length conjunction of conditions with an associated class prediction (see Fig. 1). In the evaluation, each condition compares a lagged value or the current value of the time series to a threshold (decision boundary). The decision lists in EC-Star have a variable number of rules and conjunctive clauses in each rule, but are limited by max decision list size. This representation is different from many other classifiers e.g., DecisionTrees, simple Decision Lists, Support Vector Machines and Logistic Regression, which require every time lagged value or an aggregate to be set as a different feature.

Furthermore, the EC-Star representation requires a specific layout of the data. The data is assembled as data packages, where each data package is a classification example. Consider two time series $x_1(t)$ and $x_2(t)$. Within each data package for each time interval $t = a$ the values of $x_1(a)$ and $x_2(a)$ are stored as columns. This is shown below in the example in Table 1. If the problem has more time series additional columns can be incorporated into the data package. Each data package is associated with a label l. The rule is evaluated for each data package and its error rates, false positive and false negatives are calculated by accumulating the discrepancy between its predicted label and the true label for the data package. Table 1 presents a rule and its prediction for a data package.

The quality of an evolved decision list (i.e., the candidate) is determined by the weighted error (WE). L is the set of labels. $C_{i|j}$ is the cost of predicting label i as j, and $p_{i|j}$ is the probability of predicting the label i when it is actually j.

$$WE = \sum_{j]inL} \sum_{i \in L} (C_{i|j} . p_{i|j}) \tag{1}$$

The cost is

Table 1 Example of a decision list (*top*) and a data package (*bottom*), and the evaluation of a rule on a data package

Number	Condition1	Condition2	Action
1	if $x_1(T) < 10$ and	$x_2(T-1) > 20$	then $l = 0$
2	if $x_2(T) < 10$ and	$x_2(T-3) > 20$	then $l = 1$
3	if $x_1(T-1) < 10$		then $l = 0$

T	$x_1(T)$	$x_2(T)$	l
0	10	20	
1	10	32	
2	9	30	
3	8	20	0

The example shows how the current Time, $T = 3$ is applied to each rule. The first rule evaluates the first clause with x_1 at the current time to true. The second clause of the first rule, x_2 at current time -1, $T - 1 = 2$ is also true. The first rule applies label $l = 0$ as the action, which matches the label for $T = 3$. The first clause in rule number 2 compares x_2 at $T = 3$, which is false. The second clause compares x_2 at $T - 3 = 0$, which is again false (this second clause does not have to evaluated given the first one is false). Thus, the rule takes no action. The third rule compares x_1 at time $T - 1 = 2$ and is true and takes the action $l = 0$, which is correct. When there is more than one prediction the current heuristic for choosing the action is to take the first prediction, in the same manner as a decision list. Thus, the action from rule number one, $l = 0$, is predicted, which is correct. If no rules are true then the action will be 'Null', which is always incorrect

$$C = \begin{pmatrix} 0 & 1 & 1 \\ 500 & 0 & 1 \\ 600 & 1 & 0 \\ 600 & 1 & 1 \end{pmatrix}$$

Each data sample includes a time-series of 100 labeled events. The quality of predictions is aggregated and normalized by the candidate's age to calculate the candidate's fitness at any given time.

We demonstrate the efficacy of the approach on roughly 4000 patients Arterial Blood Pressure (ABP) waveforms from MIMIC II v3. In MIMIC Waveform records available are sampled at 125 Hz (125 samples/s) (Goldberger et al. 2000) and ABP is recorded invasively from one of the radial arteries. The raw data size was roughly 1 TB. The labels in the data are imbalanced, the total number of Low event are just 1.9 % of the total number of events. In total we had 45,693 EC-Star data packages from 4414 patient records. Of these, we used 32,898 packages with 100 events each as the training set. 12,795 samples are used for the unseen set.

4 Experiments

For the purposes of the experiments in this paper, we split the training data into 1, 2, 4, 8, 16, 32, and 64 mutually exclusive segments respectively. The data sets are regenerated randomly from the training set before each experiment.

All experimental runs used the same number of Evolution Engine processing nodes, namely 64, with a run duration of 24 h. The maturity age, and consequently, the age-layer range, in all experiments was set to 150. The age-layer quota for all runs was set at 100 candidates per age-layer. The Evolution Engine pool size for all runs was fixed at 1000 candidates, with an elitist percentage of 10 %.

For the nPool experiments, the top-layer min age for the runs is reduced in inverse proportion to the number of segments. For n=1, top-layer min-age is 30,000. For n=2, it is set to 15,000, for n=4, it is 7500, for n=8, it is 3750, for n=16, it is 1800, for n=32, it is 900, and for 64, it is 450. In all cases, of course, the harvest age is the same (i.e., 30,000).

4.1 Results

In Fig. 4 we compare average fitness results from training. Note that the training results include the results from running on the unseen segments in the cases where *n* is greater than one. The unseen set is the same for all runs.

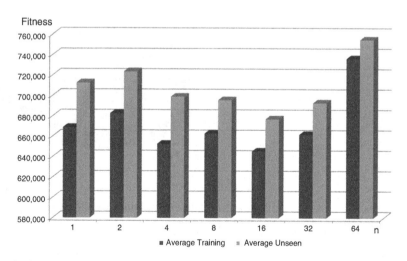

Fig. 4 Comparison of nPool training and unseen results between runs with 1 (i.e., no nPool), 2, 4, 8, 16, 32, and 64 segments for the top four candidates. These candidates are then run on the unseen set withheld from training for comparison

Fig. 5 Distribution of
originating segments for 100
fittest candidates in top-layer
of runs with 2 segments. Each
slice in chart represents a
segment

**Hundred Best Genes Distribution Among Two
Segments**

Fig. 6 Distribution of
originating segments for 100
fittest candidates in top-layer
of runs with 4 segments

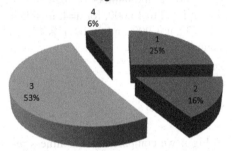

**Hundred Best Genes Distribution Among Four
Segments**

Fig. 7 Distribution of
originating segments for 100
fittest candidates in top-layer
of runs with 8 segments

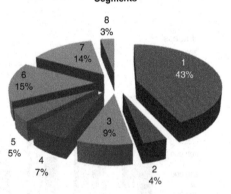

**Hundred Best Genes Distribution Among Eight
Segments**

Figures 5, 6, 7, 8, 9, and 10 show the relative contribution of different originating
segments to the top 100 candidates of each run.

Fig. 8 Distribution of originating segments for 100 fittest candidates in top-layer of runs with 16 segments. Note that segments 1, 9, and 12 have no candidates represented in the top 100

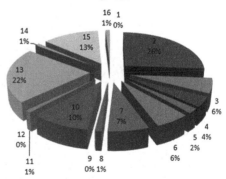

Hundred Best Genes Distribution Among Sixteen Segments

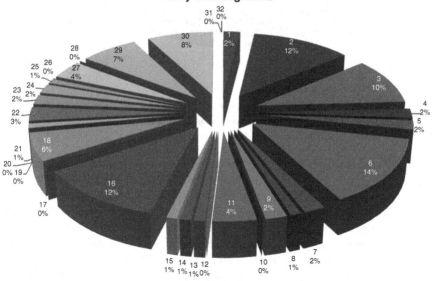

Hundred Best Genes Distribution Among Thirty Two Segments

Fig. 9 Distribution of originating segments for 100 fittest candidates in top-layer of runs with 32 segments. Note that segments 10, 12, 17, 19, 20, 26, 28, 31 and 32 have no candidates represented in the top 100

4.2 Discussion

The experiments show that running in nPool mode does not adversely impact the quality of the results, and, there's a hint that it actually may be helping to improve them. This may be due to differences in the makeup of the originating segments, allowing for better generalization. We base this on the evidence of disparity in the contribution of different originating segments to the make-up of the top performers.

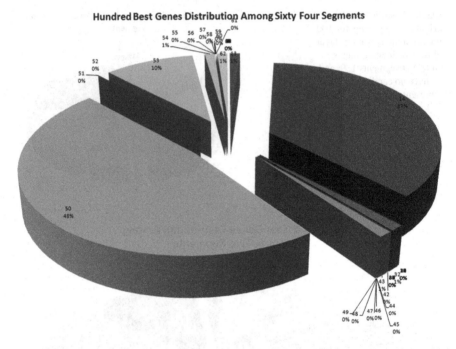

Fig. 10 Distribution of originating segments for 100 fittest candidates in top-layer of runs with 64 segments. Note that 56 of the 64 segments have no candidates represented in the top 100

In Fig. 7, for example, segment 1 has a disproportionately large contribution to the top layer. It is as if candidates that were trained on segment 1 had a better potential to learn and generalize. By the same token, we believe that the resulting candidates originating from different segments have a higher diversity. This needs to be investigated and qualified further and we hope to get to it in our future work.

The results show that, when n increases, the number of segments contributing to the top candidates drops. For example, for $n = 16$ three of the segments do not contribute at all to the top 100 candidates from the run. This number goes up to 56 for $n = 64$. This is mitigated by other segments that seem to have a better representation of the data and so allow for even better generalization and overall results. There is a point, however, after which the size of the segments, regardless of the makeup, cannot sustain the run (e.g., see the 64 segment run results in Fig. 4).

5 Conclusions

We described the nPool model for cross-validation in a distributed evolutionary system with incremental fitness evaluations. The real-world experimental results for this approach are promising. Here are some of the benefits of this method:

1. *Diversity promotion*

Training takes place on a diversity of mutually exclusive training sets, which can potentially lead to a diversity of solutions.

2. *Generalization*

The method allows for much more effective use of the data set, removing the requirement to always hide part of the data from any training to be used as the out-of-sample set. The method also significantly reduces sensitivity to any selection bias on the training set by allowing n originating segments, each to act as a training set for a subset of the candidates. This method removes, or at least reduces, the need for a separate process for verifying evolved candidates on out-of-sample data, as that step is built into the production system. In addition, generalization of any evolved candidate is much more reliable as its respective out-of-sample evaluation set is $n - 1$ times larger than the evolved set.

3. *Scale*

Many more candidates are tested on unseen, this is done in parallel, and it is simultaneous to the training run.

4. *Speed*

A lower top layer max age means faster convergence over the training sets (i.e., originating segments). However, this is evened out somewhat because more time and processing capacity is spent on validation, so less capacity is available for training. The age-layered nature of the system filters out over-fitting candidates, so segments with uneven distributions of data points have less of an impact. This manner of association of segments to Evolution Engines has the added benefit of allowing for the caching of the data points at the worker nodes, reducing the need for moving data packages around. This allows the infrastructure to be smarter about moving the candidates around rather than the data—reducing the bandwidth requirements and, as a result, improving the efficiency of the system.

5.1 Future Work

One of the drawbacks to the system just described is that the single fold of data on which the candidates are trained is fixed throughout training, with no overlap. This means that there is a chance that the training pools associated with each data segment may converge relatively quickly, and the diversity of data is not translated into a diversity of genotypic solutions. In other words, if the data segments are small enough, there is a risk that candidates with the same source segment evolve to local optima much quicker than if we trained on the entire data set.

To combat this problem, we can alter the approach so that an Evolution Coordinator designates each Evolution Engine to evaluate material from M different sub-segments, where M is less than the total number of data segments in the system (often less than $1/2$). Candidates originating from an Evolution Engine are also marked as originating from all of the segments available to the Evolution Engine.

Training continues until the maturity age for all of its M folds is reached. At this point validation can then begin on the remaining $N - M$ segments, by sending the candidate back down to Evolution Engines designated to sub-segment pair unseen by the candidate so far.

Note that M is really a maximum for a given data set. An individual Evolution Engine can be assigned any number of folds from 1 to M. By doing so, we have many more permutations of the data sub-segments assigned as originating segments, and the segments have overlaps. For instance, rather than dividing the data set to a fixed 4 segments, we can divide it to 8 sub-segments, and assign each Evolution Engine two of the possible permutations of the sub-segments, which would be a total of $8 \times 7 = 56$ possible pairs to be used as originating segments. This would mean, however, that a candidate should be prevented from being sent for evaluation to an Evolution Engine with a combination of sub-segments that include any sub-segment in the candidate's originating segment set.

More experimental and theoretical work on the approach and the best settings for n are also in order.

Acknowledgements The authors wish to thank Sentient Technologies for sponsoring this research and providing the processing capacity required for the experiments presented in this paper.

References

Goldberger AL, Amaral LA, Glass L, Hausdorff JM, Ivanov PC, Mark RG, Mietus JE, Moody GB, Peng CK, Stanley HE (2000) Physiobank, physiotoolkit, and physionet components of a new research resource for complex physiologic signals. Circulation 101(23):e215–e220

Hodjat B, Shahrzad H (2013) Introducing an age-varying fitness estimation function. In: Genetic programming theory and practice X. Springer, Berlin, pp 59–71

Hodjat B, Hemberg E, Shahrzad H, O'Reilly UM (2014) Maintenance of a long running distributed genetic programming system for solving problems requiring big data. In: Genetic programming theory and practice XI. Springer, Berlin, pp 65–83

Hornby, GS (2006) ALPS: the age-layered population structure for reducing the problem of premature convergence. In: Proceedings of the 8th annual conference on Genetic and evolutionary computation, ACM, New York, pp 815–822

O'Reilly UM, Wagy M, Hodjat B (2013) Ec-star: a massive-scale, hub and spoke, distributed genetic programming system. In: Genetic programming theory and practice X. Springer, Berlin, pp 73–85

Refaeilzadeh P, Tang L, Liu H (2009) Cross-validation. In: Encyclopedia of database systems. Springer, Berlin, pp 532–538

Rivest RL (1987) Learning decision lists. Mach Learn 2(3):229–246

Highly Accurate Symbolic Regression with Noisy Training Data

Michael F. Korns

Abstract As symbolic regression (SR) has advanced into the early stages of commercial exploitation, the poor accuracy of SR, still plaguing even the most advanced commercial packages, has become an issue for early adopters. Users expect to have the correct formula returned, especially in cases with zero noise and only one basis function with minimally complex grammar depth.

At a minimum, users expect the response surface of the SR tool to be easily understood, so that the user can know a priori on what classes of problems to expect excellent, average, or poor accuracy. Poor or *unknown* accuracy is a hindrance to greater academic and industrial acceptance of SR tools.

In two previous papers, we published a complex algorithm for modern symbolic regression which is extremely accurate for a large class of Symbolic Regression problems. The class of problems, on which SR is extremely accurate, is described in detail in these two previous papers. This algorithm is extremely accurate, in reasonable time on a single processor, for from 25 up to 3000 features (columns).

Extensive statistically correct, out of sample training and testing, demonstrated the extreme accuracy algorithm's advantages over a previously published base line pareto algorithm in case where the training and testing data contained zero noise.

While the algorithm's extreme accuracy for deep problems with a large number of features, on noiseless training data, is an impressive advance, there are many very important academic and industrial SR problems where the training data is very noisy.

In this chapter we test the extreme accuracy algorithm and compare the results with the previously published baseline pareto algorithm. Both algorithms' performance are compared on a set of complex representative problems (from 25 to 3000 features), on noiseless training, on noisy training data, and on noisy training data with range shifted testing data.

The enhanced algorithm is shown to be robust, with definite advantages over the baseline pareto algorithm, performing well even in the face of noisy training data and range shifted testing data.

M.F. Korns (✉)
Analytic Research Foundation, 2240 Village Walk Drive Suite 2305, Henderson, NV 89052, USA

Freeman Investment Management, Henderson, NV, USA
e-mail: mkorns@korns.com

© Springer International Publishing Switzerland 2016
R. Riolo et al. (eds.), *Genetic Programming Theory and Practice XIII*,
Genetic and Evolutionary Computation, DOI 10.1007/978-3-319-34223-8_6

91

Keywords Symbolic regression • Abstract expression grammars • Grammar template genetic programming • Genetic algorithms • Particle swarm

1   Introduction

The discipline of Symbolic Regression (SR) has matured significantly in the last few years. There is at least one commercial package on the market for several years http://www.rmltech.com/. There is now at least one well documented commercial symbolic regression package available for Mathematica www.evolved-analytics.com. There is at least one very well done open source symbolic regression package available for free download http://ccsl.mae.cornell.edu/eureqa. In addition to our own ARC system (Korns 2010), currently used internally for massive (million row) financial data nonlinear regressions, there are a number of other mature symbolic regression packages currently used in industry including Smits and Kotanchek (2005) and Kotanchek et al. (2008). Plus there is another commercially deployed regression package which handles up to 50–10,000 input features using specialized linear learning (McConaghy 2011).

Yet, despite the increasing sophistication of commercial SR packages, there have been serious issues with SR accuracy even on simple problems (Korns 2011). Clearly the perception of SR as a *must use* tool for important problems or as an *interesting heurism* for shedding light on some problems, will be greatly affected by the demonstrable accuracy of available SR algorithms and tools. The depth and breadth of SR adoption in industry and academia will be greatest if a very high level of accuracy can be demonstrated for SR algorithms.

In Korns (2012, 2013, 2014) we published both a baseline pareto algorithm and an extreme accuracy algorithm for modern symbolic regression the (*EA*) algorithm. which is extremely accurate for a large class of Symbolic Regression problems. The class of problems, on which the EA algorithm is extremely accurate, is described in detail in those papers and also in this chapter. A definition of extreme accuracy is provided, and an *informal argument* of extreme SR accuracy is outlined in Korns (2013, 2014).

Prior to writing this chapter, a great deal of *tinker-engineering* was performed on the Lisp code supporting both the baseline and the EA algorithms. For instance, all generated champion code was checked to make sure that the real numbers were loaded into Intel machine registers without exception. All vector pointers were checked to make sure they were loaded into Intel address registers at the start of each loop rather than re-loaded with each feature reference. As a result of these engineering efforts, both the baseline and the EA algorithms are now quite practical to run on a personal computer. Furthermore the EA algorithm is extremely accurate, in reasonable time, on a single processor, for from 25 to 3000 features (columns); and, a cloud configuration can be used to achieve the extreme accuracy performance in much shorter elapsed times.

In this chapter we test the EA algorithm (Korns 2013, 2014) and compare the results with the baseline algorithm (Korns 2012). Extensive statistically correct, out of sample training and testing, are used to compare both algorithms' performance, on a set of complex representative problems (from 25 to 3000 features), on noiseless training, on noisy training data, and on noisy training data with range shifted testing data.

The EA algorithm is shown to be robust, with definite advantages over the baseline pareto algorithm, performing well even in the face of noisy training data with range shifted testing data.

Before continuing with the comparisons of the baseline and EA algorithms, we proceed with a basic introduction to general nonlinear regression. Nonlinear regression is the mathematical problem which Symbolic Regression aspires to solve. The canonical generalization of nonlinear regression is the class of Generalized Linear Models (GLMs) as described in Nelder and Wedderburn (1972). A GLM is a linear combination of $\mathbf{I}$ basis functions B_i; $i = 0, 1, \ldots I$, a dependent variable y, and an independent data point with M features $x = <x_0, x_1, x_2, \ldots, x_{M-1}>$: such that

- *(E1)* $y = \gamma(x) = c_0 + \Sigma c_i B_i(x) + \mathbf{err}$

As a broad generalization, GLMs can represent any possible nonlinear formula. However the format of the GLM makes it amenable to existing linear regression theory and tools since the GLM model is linear on each of the basis functions B_i. For a given vector of dependent variables, Y, and a vector of independent data points, X, symbolic regression will search for a set of basis functions and coefficients which minimize **err**. In Koza (1992) the basis functions selected by symbolic regression will be formulas as in the following examples:

- *(E2)* $B_0 = x_3$
- *(E3)* $B_1 = x_1 + x_4$
- *(E4)* $B_2 = sqrt(x_2)/tan(x_5/4.56)$
- *(E5)* $B_3 = tanh(cos(x_2*.2)*cube(x_5 + abs(x_1)))$

If we are minimizing the normalized least squared error, NLSE (Korns 2012), once a suitable set of basis functions B have been selected, we can discover the proper set of coefficients C deterministically using standard univariate or multivariate regression. The value of the GLM model is that one can use standard regression techniques and theory. Viewing the problem in this fashion, we gain an important insight. Symbolic regression does not add anything to the standard techniques of regression. The value added by symbolic regression lies in its abilities as a search technique: how quickly and how accurately can SR find an optimal set of basis functions B. The immense size of the search space provides ample need for improved search techniques. In basic Koza-style tree-based Genetic Programming (Koza 1992) the genome and the individual are the same Lisp s-expression which is usually illustrated as a tree. Of course the tree-view of an s-expression is a visual aid, since a Lisp s-expression is normally a list which is a special Lisp data structure. Without altering or restricting basic tree-based GP in any way, we can view the individuals not as trees but instead as s-expressions such as this depth 2 binary tree s-exp: $(/ \ (+ \ x_2 \ 3.45)(* \ x_0 \ x_2))$, or this depth 2 irregular tree s-exp: $(/ \ (+ \ x_4 \ 3.45) \ 2.0)$.

In basic GP, applied to symbolic regression, the non-terminal nodes are all operators (implemented as Lisp function calls), and the terminal nodes are always either real number constants or features. The maximum depth of a GP individual is limited by the available computational resources; but, it is standard practice to limit the maximum depth of a GP individual to some manageable limit at the start of a symbolic regression run.

Given any selected maximum depth k, it is an easy process to construct a maximal binary tree s-expression U_k, which can be produced by the GP system without violating the selected maximum depth limit. As long as we are reminded that each f represents a function node while each t represents a terminal node (either a feature v or a real number constant c), the construction algorithm is simple and recursive as follows.

- (U_0): t
- (U_1): (f t t)
- (U_2): (f (f t t) (f t t))
- (U_3): (f (f (f t t) (f t t)) (f (f t t) (f t t)))
- (U_k): (f U_{k-1} U_{k-1})

The basic GP symbolic regression system (Koza 1992) contains a set of functions F, and a set of terminals T. If we let $t \in T$, and $f \in F \cup \xi$, where $\xi(a, b) = \xi(a) = a$, then any basis function produced by the basic GP system will be represented by at least one element of U_k. Adding the ξ function allows U_k to express all possible basis functions generated by the basic GP system *to a depth of k*. **Note to the reader**, the ξ function performs the job of a pass-through function. The ξ function allows a *fixed-maximal-depth* expression in U_k to express trees of varying depth, such as might be produced from a GP system. For instance, the varying depth GP expression $x_2 + (x_3 - x_5) = \xi(x_2, 0.0) + (x_3 - x_5) = +(\xi(x_2 \ 0.0) - (x_3 \ x_5))$ which is a *fixed-maximal-depth* expression in U_2.

In addition to the special pass through function ξ, in our system we also make additional slight alterations to improve coverage, reduce unwanted errors, and restrict results from wandering into the complex number range. All unary functions, such as *cos*, are extended to ignore any extra arguments so that, for all unary functions, $cos(a, b) = cos(a)$. The *sqroot* and *ln* functions are extended for negative arguments so that $sqroot(a) = sqroot(abs(a))$ and $ln(a) = ln(abs(a))$.

Given this formalism of the search space, it is easy to compute the size of the search space, and it is easy to see that the search space is huge even for rather simple basis functions. For our use in this chapter the function set will be the following functions: F = (+ − * / abs inv cos sin tan tanh sqroot square cube quart exp ln ξ) (where **inv**(x) = $1.0/x$). The terminal set is the features x_0 through x_{M-1} and the real constant **c**, which we shall consider to be 2^{18} in size.

Our core assertion in this chapter is that the enhanced EA algorithm will achieve, on a laptop computer, in reasonable time, extremely accurate champions for all of the problems in $U_2(1)[25]$, $U_1(25)[25]$, $U_1(5)[150]$, and in $F_{(x)}(5)[3000]$ (*note:*

$F_{(x)} = \xi$ *inv abs sqroot square cube quart exp ln cos sin tan tanh*) in reasonable computation times, of a maximum 20 h (*on an advanced laptop built in Dec 2012*) and a maximum 40 h (*on an advanced laptop built in Jan 2008*). Most noiseless problems finish far quicker than these maximum time horizons.

Pushing things to the extreme, the enhanced algorithm will achieve extremely accurate champions for all of the problems in $U_2(1)[50]$ through $U_1(5)[50]$ in a maximum of 160 h (*on an advanced laptop built in Dec 2012*). Most noiseless problems finish far quicker than these maximum time horizons.

Obviously a cloud configuration will greatly speed up the enhanced EA algorithm, and we will address cloud configurations and extreme accuracy in a later paper. For this chapter, we will develop an extremely accurate SR algorithm which any scientist can use on their personal laptop.

1.1 Example Test Problems

In this section we list the example test problems which we will address. All of these test problems lie in the domain of either $U_2(1)[25]$, $U_1(25)[25]$, $U_1(5)[150]$, or $F_{(x)}(5)[3000]$, where the function set $F_{(x)} = (\xi$ **inv abs sqroot square cube quart exp ln cos sin tan tanh**), and the terminal set is the features x_0 thru x_{M-1} plus the real number constant **c** with **cbit** = 18. Our training data sets will contain 25 features, 150, and 3000 features as specified. Our core assertion is that the enhanced algorithm will find extremely accurate champions for all of these problems and for **all similar problems** in practical time on a laptop computer.

Similar problems are easily obtained by substituting all other possibilities within $U_2(1)[25]$, $U_1(25)[25]$, $U_1(5)[150]$, or $F_{(x)}(5)[3000]$. For instance one problem in $U_2(1)[25]$ is $y = 1.687 + (94.183^*(x_3{}^*x_2))$. By substitution, $y = 1.687 + (94.183^*(x_3/x_2))$ and $y = 1.687 + (94.183^*(x_{23}{}^*x_{12}))$ are also in $U_2(1)[25]$. Another problem in $U_2(1)[25]$ is $y = -2.36 + (28.413^*\ln(x_2)/x_3)$. By substitution, $y = -2.36 + (28.413^*\cos(x_{12})^*x_6)$ and $y = -2.36 + (28.413^*\text{sqroot}(x_{21}) - x_{10})$ are also in $U_2(1)[25]$. Our core assertion is that the EA algorithm not only finds accurate solutions to the 45 test problems listed below, but also to *all other possible test problems* in $U_2(1)[25]$, $U_1(25)[25]$, $U_1(5)[150]$, or $F_{(x)}(5)[3000]$.

- **Deep problems in $U_2(1)[25]$**
- ..*Note: these problems trained on 10,000 examples with 25 features each*
- (*T1*): $y = 1.57 + (14.3^*x_3)$
- (*T2*): $y = 3.57 + (24.33/x_3)$
- (*T3*): $y = 1.687 + (94.183^*(x_3{}^*x_2))$
- (*T4*): $y = 21.37 + (41.13^*(x_3/x_2))$
- (*T5*): $y = -1.57 + (2.3^*((x_3{}^*x_0)^*x_2))$
- (*T6*): $y = 9.00 + (24.983^*((x_3{}^*x_0)^*(x_2{}^*x_4)))$
- (*T7*): $y = -71.57 + (64.3^*((x_3{}^*x_0)/x_2))$
- (*T8*): $y = 5.127 + (21.3^*((x_3{}^*x_0)/(x_2{}^*x_4)))$

- $(T9)$: $y = 11.57 + (69.113^*((x_3^*x_0)/(x_2 + x_4)))$
- $(T10)$: $y = 206.23 + (14.2^*((x_3^*x_1)/(3.821 - x_4)))$
- $(T11)$: $y = 0.23 + (19.2^*((x_3 - 83.519)/(93.821 - x_4)))$
- $(T12)$: $y = 0.283 + (64.2^*((x_3 - 33.519)/(x_0 - x_4)))$
- $(T13)$: $y = -2.3 + (1.13^*\sin(x_2))$
- $(T14)$: $y = 206.23 + (14.2^*(\exp(\cos(x_4))))$
- $(T15)$: $y = -12.3 + (2.13^*\cos(x_2^*13.526))$
- $(T16)$: $y = -12.3 + (2.13^*\tan(95.629/x_2))$
- $(T17)$: $y = -28.3 + (92.13^*\tanh(x_2^*x_4))$
- $(T18)$: $y = -222.13 + (-0.13^*\tanh(x_2/x_4))$
- $(T19)$: $y = -2.3 + (-6.13^*\sin(x_2)^*x_3)$
- $(T20)$: $y = -2.36 + (28.413^*\ln(x_2)/x_3)$
- $(T21)$: $y = 21.234 + (30.13^*\cos(x_2)^*\tan(x_4))$
- $(T22)$: $y = -2.3 + (41.93^*\cos(x_2)/\tan(x_4))$
- $(T23)$: $y = .913 + (62.13^*\ln(x_2)/\text{square}(x_4))$
- **Narrow problems in $U_1(2to3)[25]$**
- ..*Note: these problems trained on 10,000 examples with 25 features each*
- $(T24)$: $y = 13.3 + (80.23^*x_2) + (1.13^*x_3)$
- $(T25)$: $y = 18.163 + (95.173/x_2) + (1.13/x_3)$
- $(T26)$: $y = 22.3 + (62.13^*x_2) + (9.23^*\sin(x_3))$
- $(T27)$: $y = 93.43 + (71.13^*\tanh(x_3)) + (41.13^*\sin(x_3))$
- $(T28)$: $y = 36.1 + (3.13^*x_2) + (1.13^*x_3) + (2.19^*x_0)$
- **Wide problems in $U_1(5)[25]$**
- ..*Note: these problems trained on 10,000 examples with 25 features each*
- $(T29)$: $y = -9.16 + (-9.16^*x_{24}^*x_0) + (-19.56^*x_{20}^*x_{21}) + (21.87^*x_{24}^*x_2) +$
  $(-17.48^*x_{22}^*x_{23}) + (38.81^*x_{23}^*x_{24})$
- $(T30)$: $y = -9.16 + (-9.16^*x_{24}/x_0) + (-19.56^*x_{20}/x_{21}) + (21.87^*x_{24}/x_2) +$
  $(-17.48^*x_{22}/x_{23}) + (38.81^*x_{23}/x_{24})$
- **Broad problems in $F_{(x)}(5)[3000]$**
- ..*Note: these problems trained on 5000 examples with 3000 features each*
- ..*Note: $F_{(x)} = $ noop inv abs sqroot square cube quart exp ln cos sin tan tanh*
- $(T31)$: $y = 50.63 + (63.6^*\text{cube}(x_0)) + (66.54^*\text{cube}(x_1)) + (32.95^*\text{cube}(x_2)) +$
  $(4.87^*\text{cube}(x_3)) + (46.49^*\text{cube}(x_4))$
- $(T32)$: $y = -9.16 + (-9.16^*\text{square}(x_0)) + (-19.56^*\ln(x_{123})) +$
  $(21.87^*\exp(x_{254})) + (-17.48^*x_3) + (38.81^*x_{878})$
- $(T33)$: $y = 0.0 + (1^*\text{square}(x_0)) + (2^*\text{square}(x_1)) + (3^*\text{square}(x_2)) +$
  $(4^*\text{square}(x_3)) + (5^*\text{square}(x_4))$
- $(T34)$: $y = 65.86 + (79.4^*\sin(x_0)) + (45.88^*\cos(x_1)) +$
  $(2.13^*\tan(x_2)) + (4.6^*\sin(x_3)) + (61.47^*\cos(x_4))$
- $(T35)$: $y = 1.57 + (1.57/x_{923}) + (-39.34^*\sin(x_1)) +$
  $(2.13^*x_2) + (46.59^*\cos(x_{932})) + (11.54^*x_4)$
- $(T36)$: $y = 50.63 + (63.6^*\text{sqroot}(x_0)) + (66.54^*\text{sqroot}(x_1)) + (32.95^*\text{sqroot}(x_2)) +$
  $(4.87^*\text{sqroot}(x_3)) + (46.49^*\text{sqroot}(x_4))$
- $(T37)$: $y = 92.25 + (53.53^*\text{square}(2.3^*x_0)) + (88.26^*\cos(x_1)) +$
  $(42.11/x_4) + (29.0^*\text{cube}(x_3)) + (93.6^*\tanh(x_4))$

- **Broad problems in $U_1(5)[150]$**
- ..*Note: these problems trained on 10,000 examples with 150 features each*
- *(T38)*: $y = -9.16 + (-9.16^* x_{124}^* x_0) + (-19.56^* x_{120}^* x_{21}) + (21.87^* x_{24}^* x_{26}) + (-17.48^* x_{122}^* x_{23}) + (38.81^* x_{123}^* x_{24})$
- *(T39)*: $y = -9.16 + (-9.16^* x_{124}/x_0) + (-19.56^* x_{20}/x_{92}) + (21.87^* x_{102}/x_2) + (-17.48^* x_{22}/x_{143}) + (38.81^* x_{23}/x_{149})$
- *(T40)*: $y = -9.16 + (-9.16^* \cos(_0)) + (-19.56^* x_{20}/x_{21}) + (21.87^* square(x_{125})) + (-17.48^* x_{22}/x_{23}) + (38.81^* \tanh(x_{24}))$
- **Dense problems in $U_1(25)[25]$**
- ..*Note: these problems trained on 10,000 examples with 25 features each*
- *(T41)*: $y = 50.63 + (63.6^* cube(x_0)) + (66.54^* square(x_1)) + (32.95^* quart(x_2)) + (4.87^* cube(x_3)) + (46.49^* square(x_4)) + (62.85^* quart(x_5)) + (90.45^* cube(x_6)) + (63.28^* square(x_7)) + (42.15^* quart(x_8)) + (73.03^* cube(x_9)) + (92.2^* square(x_{10})) + (77.99^* quart(x_{11})) + (56.67^* cube(x_{12})) + (72.51^* square(x_{13})) + (49.77^* quart(x_{14})) + (56.94^* cube(x_{15})) + (54.76^* square(x_{16})) + (23.11^* quart(x_{17})) + (56.03^* cube(x_{18})) + (51.98^* square(x_{19})) + (11.71^* quart(x_{20})) + (33.82^* cube(x_{21})) + (46.25^* square(x_{22})) + (32.98^* quart(x_{23})) + (36.06^* cube(x_{24}))$
- *(T42)*: $y = -9.16 + (-9.16^* x_4^* x_0) + (-19.56^* x_0^* x_1) + (21.87^* x_1^* x_2) + (-17.48^* x_2^* x_3) + (38.81^* x_3^* x_4) + (3.1^* x_4^* x_5) + (59.81^* x_5^* x_6) + (93.1^* x_6^* x_7) + (.81^* x_7^* x_8) + (9.21^* x_8^* x_9) + (-5.81^* x_9^* x_{10}) + (-.01^* x_{10}^* x_{11}) + (4.21^* x_{11}^* x_{12}) + (68.81^* x_{12}^* x_{13}) + (-8.81^* x_{13}^* x_{14}) + (2.11^* x_{14}^* x_{15}) + (-7.11^* x_{15}^* x_{16}) + (-.91^* x_{16}^* x_{17}) + (20.0^* x_{17}^* x_{18}) + (1.81^* x_{18}^* x_{19}) + (9.71^* x_{19}^* x_{20}) + (8.1^* x_{20}^* x_{21}) + (6.1^* x_{21}^* x_{22}) + (18.51^* x_{22}^* x_{23}) + (7.1^* x_{23}^* x_{24})$
- *(T43)*: $y = 0.0 + (1^* square(x_0)) + (2^* square(x_1)) + (3^* square(x_2)) + (4^* square(x_3)) + (5^* square(x_4)) + (6^* square(x_5)) + (7^* square(x_6)) + (8^* square(x_7)) + (9^* square(x_8)) + (10^* square(x_9)) + (11^* square(x_{10})) + (12^* square(x_{11})) + (13^* square(x_{12})) + (14^* square(x_{13})) + (15^* square(x_{14})) + (16^* square(x_{15})) + (17^* square(x_{16})) + (18^* square(x_{17})) + (19^* square(x_{18})) + (20^* square(x_{19})) + (21^* square(x_{20})) + (22^* square(x_{21})) + (23^* square(x_{22})) + (24^* square(x_{23})) + (25^* square(x_{24}))$
- *(T44)*: $y = 65.86 + (79.4^* \sin(x_0)) + (45.88^* \cos(x_1)) + (2.13^* \tan(x_2)) + (4.6^* \sin(x_3)) + (61.47^* \cos(x_4)) + (30.64^* \tan(x_5)) + (51.95^* \sin(x_6)) + (47.83^* \cos(x_7)) + (4.21^* \tan(x_8)) + (37.84^* \sin(x_9)) + (62.57^* \cos(x_{10})) + (4.68^* \tan(x_{11})) + (32.65^* \sin(x_{12})) + (86.89^* \cos(x_{13})) + (84.79^* \tan(x_{14})) + (31.72^* \sin(x_{15})) + (90.4^* \cos(x_{16})) + (93.57^* \tan(x_{17})) + (42.18^* \sin(x_{18})) + (47.91^* \cos(x_{19})) + (41.48^* \tan(x_{20})) + (39.47^* \sin(x_{21})) + (48.44^* \cos(x_{22})) + (34.75^* \tan(x_{23})) + (56.7^* \sin(x_{24}))$
- *(T45)*: $y = 1.57 + (1.57^* x_0) + (-39.34^* \sin(x_1)) + (2.13^* x_2) + (46.59^* (x_3/x_2)) + (11.54^* x_4) + (30.64^* \ln(x_5)) + (51.95^* abs(x_6)) + (47.83^* (x_7^* x_3)) + (4.21^* quart(x_8)) + (37.84^* x_9) + (62.57^* square(x_{10})) + (4.68^* sqroot(x_{11})) + (32.65^* (x_{12}/x_3)) + (86.89^* x_{14}) + (84.79^* \tan(x_{15})) + (31.72^* cube(x_{16})) + (90.4^* (x_{17}^* x_{18})) + (93.57^* (x_{17}/x_{16})) + (42.18^* \sin(x_{18})) + (47.91^* \cos(x_{19})) + (41.48^* \ln(x_{20})) + (39.47^* square(x_{21})) + (48.44^* x_{22}) + (34.75^* (x_{23}^* x_{24})) + (56.7^* x_{24})$

2   Training with Zero Noise

Comparing the SR performance of the baseline algorithm and the EA algorithm, on noiseless training data, using statistical best practices out-of-sample testing methodology, requires the following procedure. For each sample test problem, a matrix of independent variables is filled with random numbers between -10 and $+10$. Then the specified sample test problem formula is applied to produce the dependent variable. These steps will create the training data (each matrix row is a *training example* and each matrix column is a *feature*). A symbolic regression will be run on the training data to produce the champion estimator. Next a matrix of independent variables is filled with random numbers between -10 and $+10$. Then the specified sample test problem formula is applied to produce the dependent variable. These steps will create the testing data. The fitness score is the root mean squared error divided by the standard deviation of Y, NLSE. The estimator will be evaluated against the testing data producing the final NLSE for comparison.

The baseline algorithm and the EA algorithm will be trained on each of the 45 sample test problems for comparison. The baseline algorithm halts automatically when it achieves an extremely accurate champion on the training data. The EA algorithm halts automatically when it achieves an extremely accurate champion on the training data; but the EA algorithm also halts automatically when it has exhausted it predefined search pattern. Each algorithm will be given a maximum of 20 h for completion, at which time, *if the SR has not already halted*, the SR run will be terminated and the best available candidate will be selected as the final estimator champion.

In each table of results, the **Test** column contains the identifier of the sample test problem (*T01 through T45*). The **WFFs** column contains the number of regression candidates tested before finding a solution. The **Train-Hrs** column contains the elapsed hours spent training on the training data before finding a solution. The **Train-NLSE** column contains the fitness score of the champion on the noiseless training data. The **Test-NLSE** column contains the fitness score of the champion on the noiseless testing data. The **Absolute** column contains **yes** if the resulting champion contains a set of basis functions which are algebraically equivalent to the basis functions in the specified test problem.

For the purposes of this algorithm, *extremely accurate* will be defined as any champion which achieves a normalized least squares error (NLSE) of **.0001** or less on the **noiseless testing data**. In the tables of results, in this chapter, the noiseless test results are listed under the **Test-NLSE** column header.

Obviously *extreme accuracy* is not the same as *absolute accuracy* and is therefore fragile under some conditions. Extreme accuracy will stop at the first estimator which achieves an NLSE of **0.0** on the noiseless training data, and *hope* that the estimator will achieve an NLSE of **.0001** or less on the testing data. Yes, an extremely accurate algorithm is guaranteed to find a perfect champion (*estimator training fitness of 0.0*) if there is one to be found; but, this perfect champion may or may not be the one which was used to create the testing data. For instance in the target formula $y = 1.0 + (100.0 * sin(x_0)) + (.001 * square(x_0))$ we notice that the final term $(.0001 * square(x_0))$ is less significant at low ranges of x_0; but, as the absolute magnitude of x_0 increases, the final term is increasingly significant. And, this does not even cover the many issues with problematic training data ranges and poorly behaved target formulas within those ranges. For instance, creating training data in the range -1000 to 1000 for the target formula $y = 1.0 + exp(x_2 * 34.23)$ runs into many issues where the value of **y** exceeds the range of a 64 bit IEEE real number. So as one can see the concept of *extreme accuracy* is just the beginning of the attempt to conquer the accuracy problem in SR.

For the purposes of this algorithm, ***absolutely accurate*** will be defined as any champion which contains a set of basis functions which are *algebraically equivalent* to the basis functions in the specified test problem. In the tables of results, in this chapter, the absolute accuracy results are listed under the **Absolute** column header. "Yes" indicates that the resulting champion contains a set of basis functions which are algebraically equivalent to the basis functions in the specified test problem.

As mentioned, each of the problems were trained and tested on from 25 to 3000 features as specified using out of sample testing. The allocated maximum time to complete a test problem on our laptop environment was 20 h, at which time training was automatically halted and the best champion was returned as the answer. However, most problems finished well ahead of that maximum time limit.

All timings quoted in these tables were performed on a Dell XPS L521X Intel i7 quad core laptop with 16Gig of RAM, and 1Tb of hard drive, manufactured in Dec 2012 (*our test machine*).

Note: testing a single regression champion is not cheap. At a minimum testing a single regression champion requires as many evaluations as there are training examples as well as performing a simple regression. At a maximum testing a single regression champion may require performing a much more expensive multiple regression.

The results in baseline Table 1 demonstrate only intermittent accuracy on the 45 test problems. Baseline accuracy is very good with 1, 2, or 5 features in the training data. Unfortunately, Baseline accuracy decreases rapidly as the number of features in the training data increases to 25, 100, and 3000. Furthermore, there is a great deal of overfitting as evidenced by the number of test cases with good training scores and very poor testing scores.

The baseline algorithm also suffers from bloat. This is often the reason for the baseline's frequent failure to discover the absolutely accurate formula. For instance, in test problem **T19**, the correct formula is: $y = -2.3 + (-6.13 * sin(x_2) * x_3)$. The baseline algorithm returns a champion of $y = -2.3000000000033 - (6.13 * ((0.008 * (x_3 * 125.0)) * sin(x_2))) + (0.0000000000033 * tanh(square(x_{23})))$. The first term, $(0.008 * (x_3 * 125.0))$, and the last term, $(0.0000000000033 * tanh(square(x_{23})))$, are bloat and will cause serious problems in range shifted data.

In such cases of overfitting, SR becomes deceptive. It produces tantalizing candidates which, from their training NLSE scores, look really exciting. Unfortunately, they fail miserably on the testing data.

Clearly the baseline testing results in Table 1 demonstrate an opportunity for improved accuracy.

Another serious issue with the baseline algorithm is that negative results have no explicit meaning. For example, Alice runs the baseline algorithm on a large block of data for the maximum time specified. At the conclusion of the maximum specified generations, requiring a maximum of 20 h on our laptop, no candidate with a zero NLSE (*perfect score*) is returned. The meaning of this negative result is indeterminate, as one can argue that perhaps if Alice were to run the baseline algorithm for *a few more generations* an exact candidate would be discovered.

Table 1 Baseline accuracy zero noise

Test	WFFs	Train-Hrs	Train-NLSE	Test-NLSE	Absolute
T01	2K	0.03	0.0000	0.0000	Yes
T02	2K	0.02	0.0000	0.0000	Yes
T03	2K	0.03	0.0000	0.0000	Yes
T04	11K	0.11	0.0000	0.0000	Yes
T05	812K	9.00	0.0000	0.0000	Yes
T06	1246K	20.00	0.5364	0.7727	No
T07	112K	1.29	0.0000	0.0000	Yes
T08	1221K	20.00	0.0034	0.1354	No
T09	1240K	20.00	0.0484	0.9999	No
T10	1242K	20.00	0.0185	0.9999	No
T11	1117K	20.00	0.0317	0.9999	No
T12	1414K	20.00	0.0244	0.9999	No
T13	5K	0.05	0.0000	0.0000	Yes
T14	9K	0.09	0.0000	0.0000	Yes
T15	724K	20.00	0.8540	0.9348	No
T16	884K	20.00	0.0077	0.9999	No
T17	10K	0.10	0.0000	0.0000	Yes
T18	360K	4.51	0.0000	0.0000	Yes
T19	73K	0.86	0.0000	0.0000	Yes
T20	356K	4.41	0.0000	0.0000	Yes
T21	908K	20.00	0.0560	0.0222	No
T22	908K	20.00	0.0568	0.0602	No
T23	621K	8.21	0.0000	0.9999	No
T24	5K	0.05	0.0000	0.0000	Yes
T25	77K	0.88	0.0000	0.0000	Yes
T26	17K	0.18	0.0000	0.0000	Yes
T27	79K	0.85	0.0000	0.0000	Yes
T28	10K	0.10	0.0000	0.0000	Yes
T29	870K	20.00	0.1324	0.1334	No
T30	900K	20.00	0.0290	0.0099	No

Table 1 (continued)

Test	WFFs	Train-Hrs	Train-NLSE	Test-NLSE	Absolute
T31	900K	20.00	0.2104	0.2289	No
T32	179K	8.06	0.0000	0.0000	Yes
T33	280K	20.00	0.2435	0.2398	No
T34	283K	20.00	0.2028	0.2412	No
T35	251K	20.00	0.0511	0.0540	No
T36	333K	20.00	0.4524	0.4755	No
T37	255K	11.97	0.0000	0.0000	Yes
T38	275K	20.00	0.7453	0.8026	No
T39	282K	20.00	0.0403	0.9999	No
T40	249K	20.00	0.0022	0.9999	No
T41	854K	20.00	0.0455	0.0645	No
T42	978K	20.00	0.8415	0.9999	No
T43	507K	20.00	0.3838	0.8082	No
T44	517K	20.00	0.0062	0.9999	No
T45	517K	20.00	0.0024	0.9999	No

Note1: the number of regression candidates tested before finding a solution is listed in the Well Formed Formulas (WFFs) column
Note2: the elapsed hours spent training on the training data is listed in the (Train-Hrs) column
Note3: the fitness score of the champion on the noiseless training data is listed in the (Train-NLSE) column
Note4: the fitness score of the champion on the noiseless testing data is listed in the (Test-NLSE) column with .3551 average fitness
Note5: the absolute accuracy of the SR is given in the (Absolute) column with 19 absolutely accurate

Significantly, the EA results in Table 2 demonstrate extreme accuracy on the 45 test problems. This extreme accuracy is robust even in the face of problems with large number of features. More importantly, the EA algorithm achieved a perfect score on absolute accuracy. In the case of all 45 test problems, the EA algorithm was consistently absolutely accurate.

Notice the extreme search efficiency which Table 2 demonstrates. Our assertion is that the EA algorithm is getting the same accuracy on $U_2(1)[25]$, $U_1(25)[25]$, $U_1(5)[150]$, and $F_{(x)}(5)[3000]$ as if each and every single element of those sets were searched serially; and yet we are never testing more than a few million regression candidates.

Another very important benefit of extreme accuracy will only be fully realized when all undiscovered errors are worked out of our *informal argument for extreme accuracy* and when our informal argument is crafted into a complete, peer reviewed, well accepted, formal mathematical proof of accuracy. Once this goal is achieved, we can begin to make **modus tollens** arguments from negative results!

For example, our future Alice runs the EA algorithm on a large block of data for the maximum time specified. At the conclusion of the maximum time of 20 h on

Table 2 Extreme accuracy zero noise

Test	WFFs	Train-Hrs	Train-NLSE	Test-NLSE	Absolute
T01	1K	0.01	0.0000	0.0000	Yes
T02	1K	0.01	0.0000	0.0000	Yes
T03	34K	0.13	0.0000	0.0000	Yes
T04	20K	0.11	0.0000	0.0000	Yes
T05	135K	0.26	0.0000	0.0000	Yes
T06	243K	0.40	0.0000	0.0000	Yes
T07	137K	0.29	0.0000	0.0000	Yes
T08	255K	0.42	0.0000	0.0000	Yes
T09	2935K	2.19	0.0000	0.0000	Yes
T10	5087K	3.94	0.0000	0.0000	Yes
T11	576K	0.69	0.0000	0.0000	Yes
T12	198K	0.40	0.0000	0.0000	Yes
T13	1K	0.01	0.0000	0.0000	Yes
T14	37K	0.15	0.0000	0.0000	Yes
T15	1432K	1.31	0.0000	0.0000	Yes
T16	1963K	1.70	0.0000	0.0000	Yes
T17	3869K	3.30	0.0000	0.0000	Yes
T18	3927K	3.31	0.0000	0.0000	Yes
T19	972K	1.05	0.0000	0.0000	Yes
T20	644K	0.78	0.0000	0.0000	Yes
T21	8268K	6.96	0.0000	0.0000	Yes
T22	25365K	15.35	0.0000	0.0000	Yes
T23	25675K	15.66	0.0000	0.0000	Yes
T24	1K	0.01	0.0000	0.0000	Yes
T25	1K	0.01	0.0000	0.0000	Yes
T26	1K	0.01	0.0000	0.0000	Yes
T27	1K	0.01	0.0000	0.0000	Yes
T28	1K	0.01	0.0000	0.0000	Yes
T29	453K	0.60	0.0000	0.0000	Yes
T30	143K	0.31	0.0000	0.0000	Yes
T31	113K	2.05	0.0000	0.0000	Yes
T32	51K	1.10	0.0000	0.0000	Yes
T33	232K	3.00	0.0000	0.0000	Yes
T34	1471K	13.47	0.0000	0.0000	Yes
T35	715K	7.36	0.0000	0.0000	Yes
T36	139K	2.44	0.0000	0.0000	Yes
T37	465K	5.02	0.0000	0.0000	Yes
T38	599K	4.99	0.0000	0.0000	Yes
T39	134K	1.21	0.0000	0.0000	Yes

Table 2 (continued)

Test	WFFs	Train-Hrs	Train-NLSE	Test-NLSE	Absolute
T40	255K	2.23	0.0000	0.0000	Yes
T41	24K	0.38	0.0000	0.0000	Yes
T42	1901K	8.25	0.0000	0.0000	Yes
T43	119K	1.14	0.0000	0.0000	Yes
T44	80K	0.81	0.0000	0.0000	Yes
T45	216K	1.87	0.0000	0.0000	Yes

Note1: the number of regression candidates tested before finding a solution is listed in the Well Formed Formulas (WFFs) column
Note2: the elapsed hours spent training on the training data is listed in the (Train-Hrs) column
Note3: the fitness score of the champion on the noiseless training data is listed in the (Train-NLSE) column
Note4: the fitness score of the champion on the noiseless testing data is listed in the (Test-NLSE) column with .0000 average fitness
Note5: the absolute accuracy of the SR is given in the (Absolute) column with 45 absolutely accurate

our laptop, no candidate with a zero NLSE (*perfect score*) is returned. Referring to the published, well accepted formal mathematical proof of accuracy, Alice argues (*modus tollens*) that there exists no exact relationship between X and Y anywhere within $U_2(1)[25]$, $U_1(25)[25]$, and $U_1(5)[150]$ through $F_x(5)[3000]$.

3   Training with Noisy Data

Comparing the SR performance of the baseline algorithm and the EA algorithm, on noisy training data, using statistical best practices out-of-sample testing methodology, requires the following procedure. For each sample test problem, a matrix of independent variables is filled with random numbers between -10 and $+10$. Then the specified sample test problem formula is applied to produce the dependent variable. Then 20% noise is added to the dependent variable according to the following formula: $y = (y^*.8) + random(y^*.4)$. These steps will create the training data. A symbolic regression will be run on the training data to produce the champion estimator. Next a matrix of independent variables is filled with random numbers between -10 and $+10$. Then the specified sample test problem formula is applied to produce the dependent variable. No noise is added to the testing dependent variable. These steps will create the testing data. The fitness score is the root mean squared error divided by the standard deviation of Y, NLSE. The estimator will be evaluated against the testing data producing the final NLSE for comparison.

The baseline algorithm and the EA algorithm will be trained on each of the 45 sample test problems for comparison. Each algorithm will be given a maximum of 20 h for completion, at which time, *if the SR has not already halted*, the SR run will be terminated and the best available candidate will be selected as the final estimator champion.

In each table of results, the **Test** column contains the identifier of the sample test problem (*T01 through T45*). The **WFFs** column contains the number of regression candidates tested before finding a solution. The **Train-Hrs** column contains the elapsed hours spent training on the training data before finding a solution. The **Train-NLSE** column contains the fitness score of the champion on the noisy training data. The **Test-NLSE** column contains the fitness score of the champion on the noiseless testing data. The **Absolute** column contains **yes** if the resulting champion contains a set of basis functions which are algebraically equivalent to the basis functions in the specified test problem.

The added training noise causes many problems. Even *absolute accuracy* is somewhat fragile under noisy training conditions. For instance in case of the target formula $y = 1.0 + (100.0^* sin(x_0))$, the SR will be considered *absolutely accurate* if the resulting champion, after training, is the formula $sin(x_0)$. Clearly a champion of $sin(x_0)$ will always achieve a zero NLSE on noiseless testing data, but only *if trained on noiseless training data*. If a champion of $sin(x_0)$ is trained on noisy training data, the regression coefficients will almost always be slightly off and the champion will NOT achieve a zero NLSE even on noiseless testing data. So even an absolutely accurate champion (*containing the correct basis functions*) may not achieve extreme accuracy on noiseless testing data because the coefficients will have be slightly off due to the noise in the training data.

Since we have introduced 20 % noise into the training data, we do not expect to achieve *extremely accurate* results on the noiseless testing data. However, we can hope to achieve *highly accurate* results on the testing data. For the purposes of this chapter, **highly accurate** will be defined as any champion which achieves a normalized least squares error (NLSE) of **.2** or less on the **noiseless testing data**. In the tables of results, in this chapter, the noiseless test results are listed under the **Test-NLSE** column header.

The random noise added is normally distributed and symmetric (*as normally distributed as the **random** function can achieve*). The study of asymmetric noise and non-normally distributed noise will be left to another paper.

The results in baseline Table 3 demonstrate only very intermittent accuracy on the 45 test problems. Baseline accuracy is fragile in the face of training noise. High accuracy on the noiseless testing data is infrequently achieved in 12 of the 45 test problems. Absolute accuracy on the noiseless testing data is rarely achieved in 2 of the 45 test problems. There is a great deal of overfitting as evidenced by the number of test cases with good training scores and very poor testing scores. Furthermore, there is a great deal of bloat which is why absolute accuracy is rarely achieved (*i.e. the baseline algorithm rarely discovers the correct target formula*).

Significantly, the EA results in Table 4 consistently demonstrate high accuracy in 40 of the 45 test problems. Noteably, the EA algorithm does achieve frequent

Table 3 Baseline accuracy 20 % noise

Test	WFFs	Train-Hrs	Train-NLSE	Test-NLSE	Absolute
T01	1366K	20.00	0.0355	0.0187	Yes
T02	1274K	20.00	0.0010	0.1100	No
T03	1142K	20.00	0.0991	0.9920	No
T04	1284K	20.00	0.0005	0.0536	Yes
T05	1155K	20.00	0.0807	0.9999	No
T06	1201K	20.00	0.7324	0.9999	No
T07	1181K	20.00	0.0119	0.9999	No
T08	1214K	20.00	0.0017	0.9999	No
T09	1308K	20.00	0.0448	0.9999	No
T10	1210K	20.00	0.0152	0.9999	No
T11	1230K	20.00	0.0124	0.9999	No
T12	1286K	20.00	0.0189	0.9999	No
T13	1292K	20.00	0.2973	0.4052	No
T14	1242K	20.00	0.8214	0.9999	No
T15	1135K	20.00	0.8849	0.9999	No
T16	1196K	20.00	0.0370	0.9999	No
T17	1230K	20.00	0.1125	0.1339	No
T18	1057K	20.00	0.8900	0.9999	No
T19	1163K	20.00	0.1059	0.0382	No
T20	1227K	20.00	0.0002	0.1992	No
T21	1040K	20.00	0.0120	0.8882	No
T22	934K	20.00	0.0007	0.2953	No
T23	1132K	20.00	0.0001	0.9999	No
T24	1141K	20.00	0.1061	0.1734	No
T25	1054K	20.00	0.0010	0.0657	No
T26	1068K	20.00	0.1070	0.9999	No
T27	1087K	20.00	0.1555	0.9999	No
T28	1112K	20.00	0.2023	0.9999	No
T29	972K	20.00	0.6108	0.9961	No
T30	921K	20.00	0.0115	0.9999	No
T31	247K	20.00	0.1200	0.0607	No
T32	259K	20.00	0.0716	0.0148	No
T33	265K	20.00	0.3946	0.3038	No
T34	288K	20.00	0.3975	0.9999	No
T35	273K	20.00	0.3073	0.8116	No
T36	248K	20.00	0.6486	0.5438	No
T37	309K	20.00	0.1196	0.0677	No
T38	1578K	20.00	0.6697	0.6780	No
T39	1034K	20.00	0.0215	0.9952	No

Table 3 (continued)

Test	WFFs	Train-Hrs	Train-NLSE	Test-NLSE	Absolute
T40	1505K	20.00	0.1097	0.2328	No
T41	590K	20.00	0.2731	0.2731	No
T42	694K	20.00	0.3174	0.3327	No
T43	780K	20.00	0.4356	0.9263	No
T44	800K	20.00	0.6293	0.8469	No
T45	814K	20.00	0.1069	0.0717	No

Note1: *the number of regression candidates tested before finding a solution is listed in the Well Formed Formulas (WFFs) column*
Note2: *the elapsed hours spent training on the training data is listed in the (Train-Hrs) column*
Note3: *the fitness score of the champion on the noisy training data is listed in the (Train-NLSE) column*
Note4: *the fitness score of the champion on the noiseless testing data is listed in the (Test-NLSE) column with **.6339 average fitness***
Note5: *the absolute accuracy of the SR is given in the (Absolute) column with **2 absolutely accurate***

Table 4 Extreme accuracy 20 % noise

Test	WFFs	Train-Hrs	Train-NLSE	Test-NLSE	Absolute
T01	26861K	19.38	0.0993	0.0059	Yes
T02	26897K	19.61	0.0003	0.0000	Yes
T03	26922K	19.78	0.1014	0.0133	Yes
T04	26910K	18.84	0.0004	0.0000	Yes
T05	26877K	18.61	0.0948	0.0000	Yes
T06	26922K	18.83	0.1157	0.0000	Yes
T07	26948K	18.98	0.0025	0.0000	Yes
T08	26982K	19.98	0.0009	0.0000	Yes
T09	26897K	20.11	0.0176	0.0000	Yes
T10	26877K	19.33	0.0129	0.2877	Yes
T11	26924K	20.28	0.6912	0.0747	No
T12	26879K	19.73	0.0043	0.0185	No
T13	26907K	19.99	0.3199	0.0000	Yes
T14	26896K	19.96	0.8487	0.2350	No
T15	26930K	20.16	0.6712	0.1581	Yes
T16	26870K	21.84	0.0119	0.4315	Yes
T17	26949K	21.88	0.1227	0.0000	Yes
T18	26865K	21.93	0.8763	0.9999	No
T19	26896K	22.06	0.1085	0.0000	Yes
T20	26878K	22.73	0.0007	0.0668	No
T21	26983K	23.44	0.0013	0.0000	Yes
T22	26886K	22.74	0.0027	0.0000	Yes
T23	26918K	20.46	0.0006	0.0000	Yes

Table 4 (continued)

Test	WFFs	Train-Hrs	Train-NLSE	Test-NLSE	Absolute
T24	26936K	20.23	0.1057	0.0140	No
T25	26866K	19.83	0.0009	0.0157	No
T26	26941K	16.64	0.1074	0.0178	No
T27	26884K	19.27	0.1600	0.0000	No
T28	26908K	16.40	0.2059	0.0000	No
T29	26898K	16.33	0.1168	0.0000	Yes
T30	26866K	16.13	0.0036	0.0000	Yes
T31	969K	7.06	0.1084	0.0000	Yes
T32	1472K	10.63	0.0739	0.0050	No
T33	1159K	8.33	0.2726	0.0000	Yes
T34	1123K	8.17	0.0803	0.0000	Yes
T35	1038K	7.49	0.0678	0.0000	Yes
T36	1089K	8.10	0.5901	0.1083	Yes
T37	1031K	7.55	0.1186	0.0124	No
T38	1189K	6.94	0.1128	0.0000	Yes
T39	1279K	7.82	0.0426	0.0000	Yes
T40	1299K	7.72	0.0732	0.0053	No
T41	28313K	31.2	0.1947	0.0730	No
T42	29246K	41.43	0.1002	0.0534	No
T43	28079K	28.21	0.4036	0.3682	No
T44	28605K	34.88	0.0068	0.0000	No
T45	28385K	32.31	0.0375	0.1803	No

Note1: the number of regression candidates tested before finding a solution is listed in the Well Formed Formulas (WFFs) column
Note2: the elapsed hours spent training on the training data is listed in the (Train-Hrs) column
Note3: the fitness score of the champion on the noisy training data is listed in the (Train-NLSE) column
*Note4: the fitness score of the champion on the noiseless testing data is listed in the (Test-NLSE) column with **.0698 average fitness***
*Note5: the absolute accuracy of the SR is given in the (Absolute) column with **27 absolutely accurate***

absolute accuracy, even in the face of the noisy training data, in 27 of the 45 test problems. This absolute accuracy is robust even in the face of problems with large number of features (*i.e. the EA algorithm frequently discovers the correct target formula*).

Notice the EA's failure to achieve high accuracy in **TestCaseT10**. Even though the EA discovered the absolute accurate basis function, the noisy training data caused the coefficients to be seriously skewed. Additionally, the EA's problem with absolute accuracy in **TestCaseT12** is a case in point. Noteably, the EA algorithm actually does discover the absolute answer; but, on the noisy training data, the final fitness score of the correct answer is worse than the final fitness score of a multivariable formula (*containing the correct formula*). Faced with this better fitness score, the EA chooses the incorrect answer as its primary choice and the correct answer as a secondary choice. The EA has no way of discerning that the added noise has so seriously altered the training landscape.

Nevertheless, even with all these issues, the EA algorithm achieves a level of accuracy and search efficiency which raises SR to new level of performance on noisy training data (Tables 5 and 6).

Table 5 Baseline accuracy range shifting

Test	WFFs	Train-Hrs	Train-NLSE	Test-NLSE	Absolute
T01	1666K	20.00	0.2314	0.0251	Yes
T02	1742K	20.00	0.0356	0.0007	No
T03	1675K	20.00	0.1467	0.0393	No
T04	1757K	20.00	0.0413	0.0003	No
T05	1491K	20.00	0.3619	0.0889	No
T06	1785K	20.00	0.5413	0.4399	No
T07	1896K	20.00	0.0377	0.0579	No
T08	1832K	20.00	0.0336	0.1600	No
T09	1619K	20.00	0.3800	0.9998	No
T10	1655K	20.00	0.8966	0.9998	No
T11	1765K	20.00	0.8836	0.9999	No
T12	1653K	20.00	0.0017	0.0106	No
T13	1788K	20.00	0.5727	0.0824	No
T14	1808K	20.00	0.8814	0.7037	No
T15	1857K	20.00	0.6017	0.0477	No
T16	1426K	20.00	0.0117	0.9999	No
T17	1749K	20.00	0.1115	0.0540	No
T18	1681K	20.00	0.8853	0.9999	No
T19	1770K	20.00	0.3066	0.5472	No
T20	1381K	20.00	0.0011	0.0006	No
T21	1811K	20.00	0.3383	0.1484	No
T22	1838K	20.00	0.0453	0.1881	No
T23	1732K	20.00	0.0000	0.0000	No
T24	1831K	20.00	0.2500	0.0324	No
T25	1884K	20.00	0.0428	0.4828	No
T26	1686K	20.00	0.3089	0.1358	No
T27	1613K	20.00	0.4922	0.0715	No
T28	1468K	20.00	0.8754	0.8360	No
T29	1726K	20.00	0.2747	0.3448	No
T30	1638K	20.00	0.0070	0.5507	No
T31	448K	20.00	0.3765	0.3907	No
T32	453K	20.00	0.2953	0.2615	No

Table 5 (continued)

Test	WFFs	Train-Hrs	Train-NLSE	Test-NLSE	Absolute
T33	462K	20.00	0.2783	0.1566	No
T34	387K	20.00	0.6321	0.1958	No
T35	534K	20.00	0.1813	0.0617	No
T36	460K	20.00	0.6561	0.2358	No
T37	518K	20.00	0.0974	0.0124	No
T38	1759K	20.00	0.3503	0.4808	No
T39	1734K	20.00	0.0224	0.2714	No
T40	1633K	20.00	0.0124	0.2066	No
T41	571K	20.00	0.4867	0.3647	No
T42	597K	20.00	0.3211	0.3328	No
T43	599K	20.00	0.4478	0.2434	No
T44	635K	20.00	0.6385	0.8469	No
T45	741K	20.00	0.0514	0.9999	No

Note1: the number of regression candidates tested before finding a solution is listed in the Well Formed Formulas (WFFs) column
Note2: the elapsed hours spent training on the training data is listed in the (Train-Hrs) column
Note3: the fitness score of the champion on the noisy training data is listed in the (Train-NLSE) column
*Note4: the fitness score of the champion on the noiseless testing data is listed in the (Test-NLSE) column with .3357 **average fitness***
*Note5: the absolute accuracy of the SR is given in the (Absolute) column with **1 absolutely accurate***

Table 6 Extreme accuracy range shifting

Test	WFFs	Train-Hrs	Train-NLSE	Test-NLSE	Absolute
T01	26912K	13.23	0.2308	0.0000	Yes
T02	26832K	12.71	0.0326	0.0000	Yes
T03	26868K	13.05	0.1586	0.0000	Yes
T04	26937K	13.42	0.05426	0.0000	Yes
T05	26820K	12.7	0.3626	0.0000	Yes
T06	26884K	13.03	0.4908	0.0000	Yes
T07	26885K	12.96	0.0527	0.0616	No
T08	26908K	12.21	0.0621	0.0000	Yes
T09	26880K	11.85	0.1686	0.0000	Yes
T10	26870K	13.04	0.8927	0.9999	No
T11	26862K	12.98	0.8969	0.9999	No
T12	26865K	12.94	0.0017	0.0008	No
T13	26905K	13.27	0.5626	0.0716	No
T14	26914K	13.30	0.8820	0.1826	No
T15	26836K	12.21	0.6426	0.0158	Yes
T16	26859K	12.80	0.0198	0.9999	No

Table 6 (continued)

Test	WFFs	Train-Hrs	Train-NLSE	Test-NLSE	Absolute
T17	26861K	12.96	0.1223	0.0000	Yes
T18	26832K	12.66	0.9050	0.9999	No
T19	26895K	13.09	0.3130	0.0635	No
T20	26981K	13.61	0.0013	0.1339	No
T21	26885K	13.06	0.3445	0.0808	No
T22	26956K	13.42	0.0449	0.0172	No
T23	26838K	11.91	0.0002	0.0000	Yes
T24	26857K	11.87	0.2541	0.0140	No
T25	26971K	13.49	0.0425	0.0031	No
T26	26984K	13.56	0.3175	0.1258	No
T27	26892K	13.06	0.4942	0.0942	No
T28	26871K	12.88	0.8860	0.9999	No
T29	26896K	13.03	0.1359	0.0000	Yes
T30	26882K	12.89	0.0036	0.0000	Yes
T31	969K	7.10	0.3842	0.0609	No
T32	969K	7.08	0.2776	0.0287	No
T33	975K	7.09	0.2830	0.0139	No
T34	764K	7.51	0.6563	0.0000	No
T35	1094K	8.02	0.1594	0.1042	No
T36	723K	7.01	0.6640	0.1870	No
T37	1121K	8.15	0.0992	0.0073	No
T38	1560K	8.97	0.2616	0.2316	No
T39	1228K	7.46	0.0069	0.0000	Yes
T40	1995K	14.80	0.0052	0.0066	No
T41	28632K	35.34	0.4616	0.9392	No
T42	28700K	37.84	0.2008	0.1363	No
T43	27144K	17.32	0.3983	0.2914	No
T44	27670K	24.78	0.5973	0.8458	No
T45	28694K	37.34	0.0532	0.0020	No

Note1: the number of regression candidates tested before finding a solution is listed in the Well Formed Formulas (WFFs) column

Note2: the elapsed hours spent training on the training data is listed in the (Train-Hrs) column

Note3: the fitness score of the champion on the noisy training data is listed in the (Train-NLSE) column

Note4: the fitness score of the champion on the noiseless testing data is listed in the (Test-NLSE) column with **.1937 average fitness**

Note5: the absolute accuracy of the SR is given in the (Absolute) column with **14 absolutely accurate**

4   Noisy Training with Range Shifting Testing

Comparing the SR performance of the baseline algorithm and the EA algorithm, on noisy training data with range shifted testing data, using statistical best practices out-of-sample testing methodology, requires the following procedure. For each sample test problem, a matrix of independent variables is filled with random numbers between 0 and 1. Then the specified sample test problem formula is applied to produce the dependent variable. Then 20 % noise is added to the dependent variable according to the following formula: $y = (y^*.8) + random(y^*.4)$. These steps will create the training data. A symbolic regression will be run on the training data to produce the champion estimator. Next a matrix of independent variables is filled with random numbers between -1 and 0. Then the specified sample test problem formula is applied to produce the dependent variable. No noise is added to the testing dependent variable. These steps will create the testing data. The fitness score is the root mean squared error divided by the standard deviation of Y, NLSE. The estimator will be evaluated against the testing data producing the final NLSE for comparison.

Notice the range shifted testing data. All training is performed on data between 0 and 1. The SR has never seen a negative number. Furthermore, 20 % noise is added to the dependent variable during training. Finally, the testing data is in the range -1 to 0. These are mostly negative numbers which the SR has never seen during training.

The baseline algorithm and the EA algorithm will be trained on each of the 45 sample test problems for comparison. Each algorithm will be given a maximum of 20 h for completion, at which time, *if the SR has not already halted*, the SR run will be terminated and the best available candidate will be selected as the final estimator champion.

In each table of results, the **Test** column contains the identifier of the sample test problem (*T01 through T45*). The **WFFs** column contains the number of regression candidates tested before finding a solution. The **Train-Hrs** column contains the elapsed hours spent training on the training data before finding a solution. The **Train-NLSE** column contains the fitness score of the champion on the noisy training data. The **Test-NLSE** column contains the fitness score of the champion on the noiseless testing data. The **Absolute** column contains **yes** if the resulting champion contains a set of basis functions which are algebraically equivalent to the basis functions in the specified test problem.

For the purposes of this chapter, ***extremely accurate*** will be defined as any champion which achieves a normalized least squares error (NLSE) of **.0001** or less on the **noiseless testing data**. In the table of results, at the conclusion of this chapter, the noiseless test results are listed under the **Test-NLSE** column header.

Obviously *extreme accuracy* is not the same as *absolute accuracy* and is therefore fragile under some conditions. Extreme accuracy will stop at the first estimator which achieves an NLSE of **0.0** on the noiseless training data, and *hope* that the estimator will achieve an NLSE of **.0001** or less on the testing data. Yes, an extremely accurate algorithm is guaranteed to find a perfect champion (*estimator* *training fitness of 0.0*) if there is one to be found; but, this perfect champion may or may not be the estimator which was used to create the testing data. For instance in the target formula $y = 1.0 + (100.0^* sin(x_0)) + (.001^* square(x_0))$ we notice that the final term $(.0001^* square(x_0))$ is less significant at low ranges of x_0; but, as the absolute magnitude of x_0 increases, the final term is increasingly significant. And, this does not even cover the many issues with problematic training data ranges and poorly behaved target formulas within those ranges. For instance, creating training data in the range -1000 to 1000 for the target formula $y = 1.0 + exp(x_2^* 34.23)$ runs into many issues where the value of **y** exceeds the range of a 64 bit IEEE real number. So as one can see the concept of *extreme accuracy* is just the beginning of the attempt to conquer the accuracy problem in SR.

Furthermore even *absolute accuracy* is somewhat fragile under noisy training conditions. For instance in case of the target formula $y = 1.0 + (100.0^* sin(x_0))$, the SR will be considered *absolutely accurate* if the resulting champion, after training, is the formula $sin(x_0)$. Clearly a champion of $sin(x_0)$ will always achieve a zero NLSE on noiseless testing data, but only *if trained on noiseless training data*. If a champion of $sin(x_0)$ is trained on noisy training data, the regression coefficients will almost always be slightly off and the champion will NOT achieve a zero NLSE even on noiseless testing data. So even absolute accuracy is a tricky proposition with noisy training data.

As mentioned, each of the problems were trained and tested on from 25 to 3000 features as specified using out of sample testing. The allocated maximum time to complete a test problem on our laptop environment was 20 h, at which time training was automatically halted and the best champion was returned as the answer. However, most problems finished well ahead of that maximum time limit.

All timings quoted in these tables were performed on a Dell XPS L521X Intel i7 quad core laptop with 16Gig of RAM, and 1Tb of hard drive, manufactured in Dec 2012 (*our test machine*).

Note: testing a single regression champion is not cheap. At a minimum testing a single regression champion requires as many evaluations as there are training examples as well as performing a simple regression. At a maximum testing a single regression champion may require performing a much more expensive multiple regression.

The results in baseline Table 1 demonstrate only intermittent accuracy on the 45 test problems. Baseline accuracy is very good with 1, 2, or 5 features in the training data. Unfortunately, Baseline accuracy decreases rapidly as the number of features in the training data increases to 25, 150, and 3000. Furthermore, there is a great deal of overfitting as evidenced by the number of test cases with good training scores and very poor testing scores.

In such cases of overfitting, SR becomes deceptive. It produces tantalizing candidates which, from their training NLSE scores, look really exciting. Unfortunately, they fail miserably on the testing data.

Clearly the baseline testing results in Table 1 demonstrate an opportunity for improved accuracy.

Another serious issue with the baseline algorithm is that negative results have no explicit meaning. For example, Alice runs the baseline algorithm on a large block of data for the maximum time specified. At the conclusion of the maximum specified generations, requiring a maximum of 20 h on our laptop, no candidate with a zero NLSE (*perfect score*) is returned. The meaning of this negative result is indeterminate, as one can argue that perhaps if Alice were to run the baseline algorithm for *a few more generations* an exact candidate would be discovered.

Significantly, the EA results in Table 2 demonstrate extreme accuracy on the 45 test problems. This extreme accuracy is robust even in the face of problems with large number of features.

Notice the extreme search efficiency which Table 2 demonstrates. Our assertion is that the EA algorithm is getting the same accuracy on $U_2(1)[25]$, $U_1(25)[25]$, $U_1(5)[150]$, and $F_{(x)}(5)[3000]$ as if each and every single element of those sets were searched serially; and yet we are never testing more than a few million regression candidates.

Another very important benefit of extreme accuracy will only be fully realized when all undiscovered errors are worked out of our *informal argument for extreme accuracy* and when our informal argument is crafted into a complete, peer reviewed, well accepted, formal mathematical proof of accuracy. Once this goal is achieved, we can begin to make **modus tollens** arguments from negative results!

For example, our future Alice runs the EA algorithm on a large block of data for the maximum time specified. At the conclusion of the maximum time of 20 h on our laptop, no candidate with a zero NLSE (*perfect score*) is returned. Referring to the published, well accepted formal mathematical proof of accuracy, Alice argues (*modus tollens*) that there exists no exact relationship between X and Y anywhere within $U_2(1)[25]$, $U_1(25)[25]$, and $U_1(5)[150]$ through $F_x(5)[3000]$.

5 Conclusion

In a previous paper (Korns 2011), significant accuracy issues were identified for state of the art SR systems. It is now obvious that these SR accuracy issues are due primarily to the poor surface conditions of specific subsets of the problem space. For instance, if the problem space is exceedingly choppy with little monotonicity or flat with the exception of a single point with fitness advantage, then no amount of fiddling with evolutionary parameters will address the core issue.

In Korns (2013), an EA algorithm was introduced with an informal argument asserting extreme accuracy in a number of noiseless test problems. This enhanced algorithm contains a search language and an *informal argument*, suggesting a priori, that extreme accuracy will be achieved on any single isolated problem within a broad class of basic SR problems. In Korns (2014), the EA algorithm was enhanced to include extreme accuracy on noiseless large feature test problems.

In this paper we test the enhanced EA algorithm measuring levels of extreme accuracy on problems with noisy training data, and with range shifted testing data.

The results support the view that the pursuit if high accuracy algorithms in noiseless training data also conveys distinct and measurable advantages with noisy training data and range shifted testing data. In fact, for both noiseless training data and when trained on noisy training data, then tested on range shifted testing data, the enhanced EA algorithm is measurably faster and more accurate than the baseline algorithm. This places the Extreme Accuracy algorithm in a class by itself.

The new EA algorithm introduces a hybrid view of SR in which advanced evolutionary methods are deployed in the extremely large spaces where serial search is impractical, and in which the intractable smaller spaces are first identified and then attacked either serially or with mathematical treatments. All academics and SR researchers are heartily invited into this newly opened *playground*, as a plethora of intellectual work awaits. Increasing SR's demonstrable range of extreme accuracy will require that new intractable subspaces be identified and that new mathematical treatments be devised.

Future research must explore the possibility of developing an Extreme Accuracy algorithm for the related field of symbolic multinomial classification.

Finally, to the extent that the reasoning in this *informal argument*, of extreme accuracy, gain academic and commercial acceptance, a climate of *belief* in SR can be created wherein SR is increasingly seen as a "**must have**" tool in the scientific arsenal.

Truly knowing the strength's and weaknesses of our tools is an essential step in gaining trust in their use.

About the Author

Michael Korns is a computer scientist with professional experience at IBM Research, Chief Scientist at Tymeshare Transactions, Chief Scientist at Xerox Imaging, and currently CEO of Korns Associates. His primary area of interest is in symbolic regression classification with an emphasis on investment finance applications. His most recent research has explored extreme accuracy algorithms for symbolic regression which are robust even in the face of noisy training data and range shifted testing data.

References

Hornby GS (2006) Age-layered population structure for reducing the problem of premature convergence. In: GECCO 2006: Proceedings of the 8th annual conference on genetic and evolutionary computation. ACM, New York

Korns M (2010) Abstract expression grammar symbolic regression. In: Genetic programming theory and practice VIII. Springer, New York, Kaufmann Publishers, San Francisco, CA

Korns M (2011) Accuracy in symbolic regression. In: Genetic programming theory and practice IX. Springer, New York, Kaufmann Publishers, San Francisco CA

Korns M (2012). A baseline symbolic regression algorithm. In: Genetic programming theory and practice X. Springer, New York, Kaufmann Publishers, San Francisco, CA

Korns M (2013). Extreme accuracy in symbolic regression. In: Genetic programming theory and practice XI. Springer, New York, Kaufmann Publishers, San Francisco, CA

Korns M (2014). Extremely accurate symbolic regression for large feature problems. In: Genetic programming theory and practice XII. Springer, New York, Kaufmann Publishers, San Francisco, CA

Kotanchek M, Smits G, Vladislavleva E (2008) Trustable symbolic regression models: using ensembles, interval arithmetic and pareto fronts to develop robust and trust-aware models. In: Genetic programming theory and practice V. Springer, New York

Koza JR (1992) Genetic programming: on the programming of computers by means of natural selection. The MIT Press, Cambridge, MA

Koza JR (1994) Genetic programming II: automatic discovery of reusable programs. The MIT Press, Cambridge, MA

Koza JR, Bennett FH III, Andre D, Keane MA (1999) Genetic programming III: Darwinian invention and problem solving. Morgan Kaufmann, San Francisco, CA

McConaghy T (2011) FFX: fast, Scalable, deterministic symbolic regression technology. In: Genetic programming theory and practice IX. Springer, New York

Nelder JA, Wedderburn RW (1972) Generalized linear models. J Roy Stat Soc Ser A Gen 135: 370–384

Poli R, McPhee N, Vanneshi L (2009) Analysis of the effects of elitism on bloat in linear and tree-based genetic programming. In: Genetic programming theory and practice VI. Springer, New York

Schmidt M, Lipson H (2010) Age-fitness pareto optimization. In: Genetic programming theory and practice VI. Springer, New York

Smits G, Kotanchek M (2005). Pareto-front exploitation in symbolic regression. In: Genetic programming theory and practice II. Springer, New York

Using Genetic Programming for Data Science: Lessons Learned

Steven Gustafson, Ram Narasimhan, Ravi Palla, and Aisha Yousuf

Abstract In this chapter we present a case study to demonstrate how the current state-of-the-art Genetic Programming (GP) fairs as a tool for the emerging field of Data Science. Data Science refers to the practice of extracting knowledge from data, often Big Data, to glean insights useful for predicting business, political or societal outcomes. Data Science tools are important to the practice as they allow Data Scientists to be productive and accurate. GP has many features that make it amenable as a tool for Data Science, but GP is not widely considered as a Data Science method as of yet. Thus, we performed a real-world comparison of GP with a popular Data Science method to understand its strengths and weaknesses. GP proved to find equally strong solutions, leveraged the new Big Data infrastructure, and was able to provide several benefits like direct feature importance and solution confidence. GP lacked the ability to quickly build and test models, required much more intensive computing power, and, due to its lack of commercial maturity, created some challenges for productization as well as integration with data management and visualization capabilities. The lessons learned leads to several recommendations that provide a path for future research to focus on key areas to improve GP as a Data Science tool.

Keywords Genetic programming • Data Science • Gradient boosted regression • Machine learning • Industrial applications • Real-world application • Lessons learned • Diversity • Ensembles

1 Introduction

Nearly 10 years ago, in this same book series, Castillo et al. (2004) evaluated Genetic Programming (GP) as a suitable technique for industrial systems modeling. The authors examined the state-of-the-art GP system developed within Dow

S. Gustafson (✉) • R. Palla • A. Yousuf
Knowledge Discovery Lab, GE Global Research, Niskayuna, NY, USA
e-mail: steven.gustafson@research.ge.com

R. Narasimhan
Data Science, GE Software, San Ramon, CA, USA

© Springer International Publishing Switzerland 2016
R. Riolo et al. (eds.), *Genetic Programming Theory and Practice XIII*,
Genetic and Evolutionary Computation, DOI 10.1007/978-3-319-34223-8_7

Chemical and compared against the typical statistical approaches. They found that GP had positive attributes, and where there were weaknesses, they recommended solutions. That work led to several enhancements to GP and motivated a workshop series led by this author for 6 years. Whereas that work stemmed out of GP applied to symbolic regression and modeling work, today, GP is sitting on the ledge of breaking out of the traditional machine learning communities to much wider adoption and impact as a potential Data Science tool.

In a recent O'Reilly report (Loukides 2010), the importance of data was stated as: "The future belongs to the companies and people that turn data into products". According to Dhar (2013), Data Science is about extracting knowledge from data, discovering new relationships between things in this world, their interactions, outcomes and predictors. Data Science in practice is about speed and the ability to answer meaningful questions effectively. It is about empowering analysts with a new skillset to leverage Big Data and analytics to make effective computational policy and business decisions. Data Scientists are often not computer scientists, and hence lack formal machine learning or artificial intelligence training. Data Scientists often come from the physical sciences where domain knowledge is leveraged to turn data into a meaningful product or outcome.

Genetic Programming is poised to become a significant enabler for Data Science. But it isn't today. In this article, we review a recent attempt to use GP for Data Science and discuss the lessons learned. We identified a novel result that is holding GP back, the iteration speed at which a Data Scientists can generate new results, that is not being addressed by existing work. In non-GP systems, iteration speed is primarily impacted by how fast can someone change the python/R/matlab scripts and re-run the code. But in GP, iteration speed might mean modifying the source code, it might mean building extra scripts to work data, or in some cases it might mean performing novel research into advanced topics like ensemble learning. Whereas some tools like Data Modeler (Castillo et al. 2004) are currently positioned well to become a Data Science tool, a new tool like DEAP (De Rainville et al. 2012) that is based on Python and open source may gain wider adoption. But both tools need more advanced capabilities in the core system to become a capable GP system for Data Science. This chapter describes the case study, introduces both methods, reports the outcomes, lessons learned and recommendations as to how GP could become a more effective Data Science tool.

2 Background

There are several trends that are enabling artificial intelligence technologies to increase their value and outcomes. The first trend has been well documented and publicized: Big Data. As computer storage and compute cycles became cheaper, new industries have grown and changed around the massive collections of data and algorithms that run on top of them. Companies like Google, Amazon and Netflix are good examples of this trend. The second trend started out as Semantic Web and later became Linked Data, and as the growing Internet began to connect devices, the trend of the Internet of Things (IoT) became popular. Inside industry, the IoT is being shaped as the Industrial Internet or The Internet of Everything, and also the Web of Things, among others (see Gustafson and Sheth (2014) for a brief introduction). This trend in the massive connection of devices with data and analytic systems only produces more data, for example from personal devices like iPhones and smart watches, but provides more sensors and actuators available for artificial intelligence technologies to sit between. The third trend, Data Science which is in many ways a direct result of the first two, is the increasing demand to leverage data to direct outcomes, either in businesses, health, or global/economic policies.

Data Science originally described the study of turning data into insights. However, more recently, the focus of Data Science has become the training and search for the skills required to practice Data Science effectively. Today, Data Science has emerged as a popular topic for students, a workforce reskilling opportunity, a major focus of national funding agencies, the focus of both private investors and startup companies, and outcomes delivery mechanism on top of the Big Data initiatives begun years earlier. The IoT predicts that billions of machines will be connected in the near future, and all those machines will be producing massive amounts of data, and algorithms and insights that can be gleaned from them will allow optimization, new businesses, and understanding that shapes policy and society.

As members of the General Electric Global Research center, we have participated in many Data Science related activities for finance, healthcare, aviation, oil and gas, power and water, and media. There are several consistently common activities shared across these industries in the Data Science tasks: accessing data, learning domain knowledge, and building descriptive and predictive models. Genetic Programming presents a compelling approach as it can both learn nonlinear models, is relatively easy to insert domain knowledge into, naturally produces a range of possible solutions, and finds solutions that can be further optimized, inspected and simplified. The latter characteristics, in particular, are interesting to a Data Scientist as it means the solution can be communicated to customers and engineers and they can "understand" how the data is being used. This is in comparison to a forest of decision trees or a neural network, for example.

Data Science as a discipline is usually described as the combination of several skills. Firstly, computer science skills are needed to work efficiently with data, statistical and math skills allow one to find complex patterns within data, a physical science background helps one to understand how to find and ask meaningful questions, and creativity is required to elegantly display and communicate results. Of course, very few people are highly skilled in each area, and thus Data Scientists are often teams of people working together. Data Scientists are usually embedded within industries, are measured by their efficiency to work with data and find patterns, and their ability to find and answer the big, high-valued questions. GP, in particular the Symbolic Regression branch of GP that deals with learning regression models, has particular relevance for the Data Science community. There is at least one GP software explicitly targeted at the Data Science space using symbolic regression, Eureqa (Schmidt and Lipson 2009; Dubcakova 2011).

Table 1 Data from various Google queries suggesting low popularity and adoption by Data Scientists in Data Science competitions on Kaggle.com, as a topic of research in Universities, on new Data Science courses, within the Government, and also within job market on LinkedIn

Approach	Kaggle forums	site:edu	site edu syllabus	site:gov	site:linkedin.com
Logistic regression	76	3620	551	50	20,800
Neural network	59	3200	46	34	4710
Random forest	79	773	9	19	2920
Genetic programming	1	92	17	13	497

The queries were constructed as (Data Science + logistic regression + modifier), where modifier would be site.edu for example

GP is at least 25 years old. Even from its initial days, learning models to fit data was a focus and is usually referred to as Symbolic Regression. For the past 6 years, the first author on this paper has run a workshop at the annual Genetic and Evolutionary Computation Conference on the topic of symbolic regression research and industry tools: the Symbolic Regression and Modeling Workshop. The workshop produced several interesting papers talks, led to new lines of research, and enabled new software tools to be highlighted to the community. Symbolic regression, the identification of a model, its variables, and their relationship (both linear and nonlinear) is at the heart of Data Science.

While GP has been shown to successfully solve problems in countless papers, it is still seen as an outsider in the mainstream machine learning and artificial intelligence community. As such, GP is still not widely available in many commercial analytical tools and Data Scientists have often not received any relevant training as to how to use the method. In Table 1, we performed several online searches to understand a rough idea of the popularity of GP as compared to other popular Data Science tools, namely logistic regression, neural networks, and random forest. One search was performed on Kaggle.com user forums, which is a popular place for thousands of Data Scientists competing on the Kaggle.com site in Data Science challenges to share and discuss the approaches and methods. The forums represent both a fascinating and educational look into how Data Scientists work. Table 1 shows that GP lags behind all other methods in all but one case. While one shouldn't place too much significance on the actual numbers in Table 1, it does support our belief that GP is not being considered as a Data Science tool.

2.1 Attributes of GP for Data Science

We now look at several requirements of Data Science and how GP can meet them as a Data Science tool.

1. Data Science leverages Big Data for data-driven decisions and outcomes. GP is a distributed search algorithm, and countless studies have look at better ways

to distribute the search process. Recent work has shown how GP can directly leverage Big Data effectively (Arnaldo et al. 2014; Fazenda et al. 2012).

2. Data Science extracts knowledge about a particular problem using data. GP produces easy-to-inspect solutions, which make it a particularly valuable method for Data Scientists.

3. Data Science leverages any existing knowledge to get to an answer. GP has a direct way of encoding knowledge into the algorithm, through functions, terminals, and objective functions, or indirectly through selection pressure or operators.

4. Data Science tools are used by many people not necessarily trained in machine learning. While there are some new tools that have intuitive interfaces (Schmidt and Lipson 2009; Wagner and Kronberger 2011; De Rainville et al. 2012; Smits et al. 2010), the core GP system is still quite complex with many parameters.

5. Data Science requires Data Scientists to iterate quickly on building new models, get feedback, and build more models. GP can often take a significant amount of time to setup, tune parameters, and search for good models.

6. Data Science tools integrate with other tools, particularly data management and visualization tools. Data Modeller and DEAP leverage the built-in capability of Mathematica and Python, whereas tools like FlexGP (Veeramachaneni et al. 2015) and Eureqa are standalone solutions that must provide their own implementations or leverage external tools.

7. Data Science tools need to perform relatively out-of-the-box. Approaches like Random Forest and Gradient Boosted Regression have become popular as they are robust with their default settings. In general, GP requires a lot of customization to make it perform well. Recent work looks to improve the basic performance capability of GP by combing it with other machine learning techniques (Icke and Bongard 2013). In O'Neill et al. (2010), several open issues are highlighted to further improve GP.

8. Data Science tools produce models that need to be implemented quickly for client-facing prototypes and demos. GP still exists in many stand alone environments, or requires a fair amount of tweaking the source packages.

2.2 Summary of Attributes of Data Science and GP

In Table 2, we summarize the attributes from the previous section and identify which GP capabilities are potential areas of concern in Data Science. Of the attributes of GP for Data Science, the inability to iterate and create new models quickly was the biggest concern for us. Some of the other issues, like integrating with big data infrastructure or simple to use interfaces, have seen progress with new commercial tools.

In Data Science, building the first model is usually a very informative and valuable task by forcing assumptions and data issues out into the open, and demonstrates a viable end-to-end pipeline of data to insight capability. There are two main challenges in building the first model. Firstly, the integration of the tool

Table 2 Data Science tool attributes and intersection with GP

	Data Science attribute	GP existing capability
1	Handle big data	Good (distributed compute and sampling)
2	Extract knowledge	Good (inspect solutions)
3	Encode domain knowledge	Good (code as functions and built-ins)
4	Easy to use	weak (lots of parameters, fewer commercial tools)
5	Iterate quickly	Weak (high compute time)
6	Integrate with big data infrastructure	Good (initial work demonstrates HDFS integration)
7	Good out of box performance	Weak (typically a lot of customization required)
8	Quick prototyping	Weak (can application developers do it?)

with data management and data featurization tools. That is, given one or more data files to build the model out of, the data often needs to be integrated, shaped, cleaned, and any derived values created. This data set is then used for modeling. Secondly, the problem specific evaluation method needs to be encoded to create end to end modeling capability that can then demonstrate incremental improvement in performance. Sometimes that might mean using an evaluation data set. Other times it might mean extrapolating on new data coming from a customer. Because this custom way of evaluating a solution changes, the ease at which it can be captured is important. We now look at a real-world Data Science activity to inform us on how GP met expectations as a Data Science tool.

3 Case Study: Operations Optimization

Our application area from industry is operations optimization, which is the improvement of one or more processes given a specific business objective. There are many different types of data used in an operations optimization problem. First, there are low-level sensors like temperature, pressure or vibration. Sensors like these are often measured in multiple places on a machine or in a plant. Secondly, there are sensors that provide states of inputs, like chemical mixture or composition, and states of outputs, like the results of a visual inspection system or a non-destructive testing method. Lastly, there are derived or back-calculated sensors that are often included in the operations optimization task. These values could be from a physics-based model or equation, or from some other equation or simulation, and then assigned back to the operation to assign a probable value. Figure 1 highlights these different input types and how they flow in the operations optimization problem.

In industrial operations, daily decisions are made regarding various control settings in order to maintain a targeted flow of product or output. Multiple sensors from the plant or field are used to understand the current state and predict the future. Given the various industrial systems encountered, the often extreme environments, and the failure or drift of sensors, the resulting real-time data is often very noisy

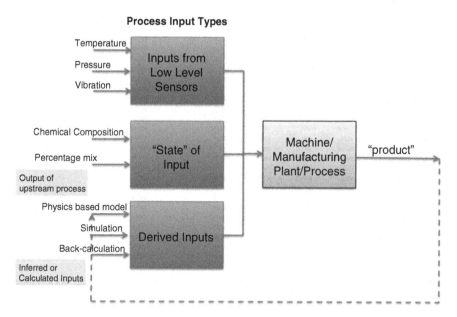

Fig. 1 Different input types to the operations optimization problem

and requires significant data processing and cleaning prior to use. Therefore, the problem at a high-level is one of using real-time sensors and control capability to understand and improve operations. Automated methods that clean up data or estimate non-measurable attributes help in providing an accurate, real-time view of the whole system. In Kordon and Smits (2001), the authors use GP to create soft sensors, or virtual sensors, that augment more expense sensors.

3.1 The Data Science Challenge at Hand

Our Data Science challenge could be stated as follows: Given historical data of an operation, is it possible to use data and analytic methods to create accurate sensor estimators, for those time instants when the sensor is offline? A sensor often changes from being online (available) or offline (unavailable). A secondary challenge was to identify instances (time periods) when the sensor of interest is drifting away from its ideal accuracy level or from its prior relationship with other system sensors. Solving these challenges would give an operations managers a consistent, real-time stream of data that characterizes their operation, enabling accurate and timely optimization decisions.

We defined our problem as having sensors $s_1 \ldots s_{N-1}$ as available, and sensor s_N as partially available: meaning that for portions of time sensor s_N is online, but it frequently goes offline due to other activities or faults. Our goal was to determine whether we could build an accurate model using data for sensors $s_1 \ldots s_{N-1}$ and

TimeStamp	Sensor 1	Sensor 2	Sensor 3	Sensor 4	Sensor 5	...	Sensor n-1	Sensor N	Online/Offline
1/1/15 0:00	value	value	value	value	value	...	value	value	Online
1/1/15 0:10	value	value	value	value	value	...	value	value	Online
1/1/15 0:20	value	value	value	value	value	...	value	value	Online
1/1/15 0:30	value	value	value	value	value	...	value	value	Online
1/1/15 0:40	value	value	value	value	value	...	value	value	Online
1/1/15 0:50	value	value	value	value	value	...	value	value	Online
1/1/15 1:00	value	value	value	value	value	...	value	value	Online
1/1/15 1:10	value	value	value	value	value	...	value	??	Offline
1/1/15 1:20	value	value	value	value	value	...	value	??	Offline
1/1/15 1:30	value	value	value	value	value	...	value	??	Offline
1/1/15 1:40	value	value	value	value	value	...	value	??	Offline
1/1/15 1:50	value	value	value	value	value	...	value	??	Offline
1/1/15 2:00	value	value	value	value	value	...	value	value	Online
1/1/15 2:10	value	value	value	value	value	...	value	value	Online
1/1/15 2:20	value	value	value	value	value	...	value	value	Online
1/1/15 2:30	value	value	value	value	value	...	value	value	Online
1/1/15 2:40	value	value	value	value	value	...	value	value	Online
1/1/15 2:50	value	value	value	value	value	...	value	value	Online
1/1/15 3:00	value	value	value	value	value	...	value	value	Online
1/1/15 3:10	value	value	value	value	value	...	value	??	Offline
1/1/15 3:20	value	value	value	value	value	...	value	??	Offline
1/1/15 3:30	value	value	value	value	value	...	value	??	Offline
1/1/15 3:40	value	value	value	value	value	...	value	??	Offline
1/1/15 3:50	value	value	value	value	value	...	value	??	Offline

Fig. 2 Schematic illustrating the input Time Series Data, with one sensor that goes offline periodically

data from sensor s_N (when it was available). An operation has many moving parts and changing system dynamics over time will cause the sensors to drift from their relationships to each other and various business objectives. In our case, we had data for 1 year, with measurements consistently every 10 min for all sensors and a binary value to tell us whether sensor s_N was offline or online. A conceptual schematic of the input data is shown in Fig. 2.

As part of the Data Science study, two different but related methods were used. The first method, Gradient Boosted Regression (GBR), has been commonly and successfully employed in multiple Data Science competitions and is available in several open source packages. The other method is GP for Symbolic Regression. In particular, a system developed by MIT which is referred to in Arnaldo et al. (2014) as a competent GP, because it contains many state-of-the-art features. Since this was an actual Data Science engagement, with real data and a client waiting for the results, and not a simulated experiment for publication, we could only attempt two different methods given our deadlines and commitments. To measure and compare the performance of the two methods, we used the Root Mean Squared Error (RMSE) over the period of extrapolation as a measure of accuracy.

As mentioned before, sensor s_N is available (online) only during certain time intervals and those periods in the historical data constitute our training and testing data. On this dataset, with the power of either of the two techniques being compared, getting very low RMSE was easy if we resorted to interpolation. In other words, if we used input data that sandwiched the time period of interest (from both the past and the future) to predict the value of sensor s_N, we obtain good predictions. This is due to the fact that there exist very strong temporal relationships within the data. Therefore, creating a model using training data that spans the time period of prediction (and withhold the test points from within that time period), can accurately interpolate the test data. However, this would not be a true test, since in actual implementation, future values would never be available to use. We would only have historical data for model building. In order to address this, and to truly assess the capabilities of the two methods, we limited both methods to use only training data that occurred prior in time when making any predictions. Therefore, our test measure of RMSE is technically a measure of extrapolation (forward looking), as opposed to interpolation. How do we know if the RMSE that we obtained is really good, or just good enough? To guide us for this, we can use the raw data variability from sensor s_N. Ideally, the RMSE and the raw data variability should be within the same magnitude. In our study, using six different periods of testing, we had an average standard deviation of actual sensor s_N value (a percentage) to be around 3.5. The GP system was able to achieve an average RMSE over the same six periods of 6.0, and the GBR system had 5.5. Thus, both approaches achieved reasonable extrapolation capabilities.

3.2   Data Management

One aspect that is related to Data Science but often not mentioned in the GP literature is the common task of managing data. In academic settings, artificial intelligence and machine learning research is often carried out using pre-cleaned or benchmark data sets. However, in industry, the challenge of gathering, organizing, and preparing data is significant. Particularly when collaborating between multiple people using different approaches, data management done poorly can lead to significant issues, and in some cases call into question the validity of results. In our work, we used an approach that is growing interest in industry: ontology-based data access. We created a model of the domain, link our data to that domain model, and then employ a suitable query system to shape and access data. Given that the data we have corresponds to sensor readings, we create models of the different types of things that have sensors, their properties, and the relationships between those things. For example, an electric submersible pump may be part of an oil well, or a high pressure turbine blade may be a part of a gas turbine, and a pump may have some rotation frequency. The sensor readings typically correspond to the properties like rotation frequency.

We used the Web Ontology Language (OWL), which is a W3C standard for representing ontologies, to capture the domain model. Figure 3 (generated using the OntoGraf plugin in Protégé[1]) shows a sample ontology describing a part of our system.

---

[1]http://protege.stanford.edu/

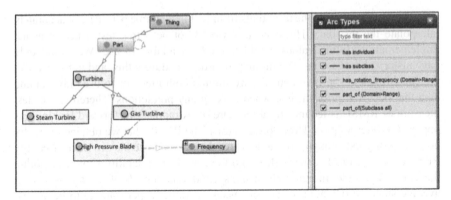

Fig. 3 A sample of the ontology from our system, showing how a gas turbine is a part, as well as has a part (compressor), which itself has a part (high pressure turbine blade), that has a sensor which measures rotation frequency

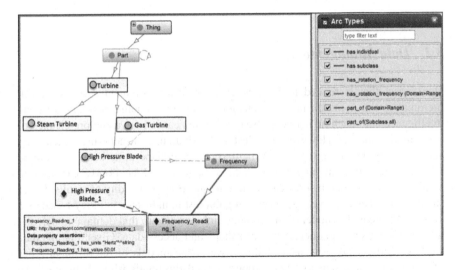

Fig. 4 The ontology of the system with frequency data added for the frequency concept

With the ontology in place, we then map the sensor data to the domain model by representing the data as properties of the instances of the corresponding components. Figure 4 extends the above figure by adding frequency data.

This approach to data management provides an intuitive representation of the sensor data since it maps the data to the domain model. The queries used to retrieve the data are also based on the domain model, thus making it easier for different users to understand and modify the queries. We use SPARQL (SPARQL Protocol and RDF Query Language), which is a W3C standard for querying the domain model and data we represent in RDF format (Resource Description Framework). SPARQL queries were saved and shared between the team to allow easy communication of the data sets and preparation. In a recent edition of AI Magazine, there are several articles highlighting the value of Semantic approaches for data management (van Harmelen et al. 2015).

3.3 Gradient Boosted Regression

We needed another method to compare against and for this we focused on a popular Data Science technique called Gradient Boosted Regression (GBR). GBR can be used to build predictive models for regression problems. GBR was first proposed in Friedman (2001). In brief, GBR uses a weighted ensemble of very simple classification and regression trees (CART) to estimate a function. After creating an initial model, GBR then uses gradient descent to find subsequent models that minimize the errors of the residuals from the previously selected model. This technique has emerged to become a very powerful and efficient technique for regression as well as classification problems.

GBR has a way to combine several shallow trees (called learners). Each tree by itself is fairly weak in its ability to predict, but when combined they 'grow' into a strong predictor. These trees are iteratively modified, based on the residuals, or model error, the difference between the predicted values and the actuals. In GBR, the goal is to minimize that difference, or residual, measured by a loss function. In gradient boosting, we start with an imperfect model, and stage by stage try to improve it. Each improvement is informed by the residual of the preceding imperfect model. It turns out that the residuals are the negative gradients of the loss function (hence the name). In summary, in GBR, we successively reduce the loss function and thereby generating a better set of parameters. By using a selection of freely available Python libraries, we were able to very easily import the data, format it to feed it to the model for training, testing and validation. Specifically, we used Pandas for data manipulation, and the SciP, NumPy, and Scikit-Learn Python modules for GBR itself (Jones et al. 2001; van der Walt et al. 2011; Pedregosa et al. 2011).

3.4 Practical Considerations When Implementing the Gradient Boosting Method

We created the initial GBR scripts in Python. Note that there are excellent and well-documented GBR implementations available in Python, Matlab and R. The initial solution script was fairly straightforward to build. Once the experiment setup was complete (splitting of data into training and testing, and a way to cross validate), there were numerous online resources available to demonstrate how to easily use the multiple Python libraries to build a GBR model and to tune its parameters.

The ways we improved the GBR results were through feature selection, by increasing the training data, and by re-training on a sliding time-window. Once we started to obtain acceptable results, the default parameters used for GBR were not changed. The Python GBR approach was very quick to build, and we used common and open-source libraries to create the initial Python script. In total, around 400 lines of un-optimized Python code were written to implement the complete GBR solution.

3.5   Adapting the Model to Handle Sensor Drift

One effect of implementing a predictive model that relies on time series data as its input is that the model's accuracy will tend to degrade over time as sensor values drift. This is because the underlying relationships between the sensors $s_1 \ldots s_N$ that were modeled have themselves changed over time. If we do not account for this, a model that was accurate at a previous point in time will start to make very poor predictions. The typical way to fix this is to 'rebuild' the model by re-estimating the predictors. This raises two questions:

1. How do we know when it is time to refresh the existing model?
2. How frequently should we 'publish' a new model to the clients such that it causes minimal disruption?

To prevent this error from growing large over time, we ran two models in parallel. One model was the existing (published) model, and the other was a new one being evaluated. In addition to the comparing the prediction errors, we also compared the actual predictor values and the variable ranks of the two models. If the updated model was 'significantly' different (above the set threshold), it was accepted as time to switch to the new model and publish it to the clients (Fig. 5).

3.6   Genetic Programming Solution

Genetic Programming is an evolutionary algorithm which provides a heuristic search alternative to other modeling approaches such as regression (Koza 1992). A recent special issue provides an up-to-date account of progress and open issues in the field (O'Neill et al. 2010).

The GP system we used for symbolic regression contains several state-of-the-art features like multi-objective optimization for selecting for solution accuracy and complexity the maintenance of a pareto front. The underlying GP system is the same as described in Veeramachaneni et al. (2015) where additional features for cloud-based implementation, multiple-regression learning, and factoring—each learner or individual leverage a different set of parameters and data when learning. The Java-based system contained several classes implementing specific features

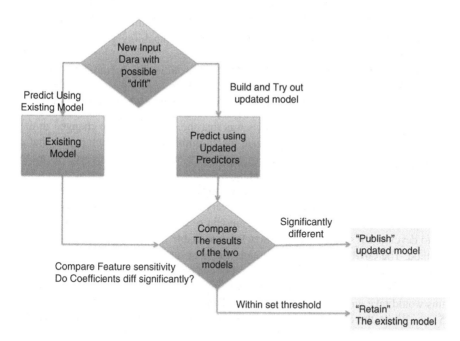

Fig. 5 Schematic showing criteria for handling sensor "Drift"

like selection, evaluation, and recombination. The GP system also contains several configuration files, or parameters file, where things like population size, functions, initialization method and selection pressure can be specified. The goal of the configuration files are to allow the customization of the system without modifying the class files, which would require a recompile of the source files. Like most systems for EC, there is a decent learning curve to understand how certain functionality is represented and programmed.

If we look across all existing GP solutions, each provides strengths in various attributes: user interfaces, cloud and distributed compute support, integration with data management and visualization solutions like Mathematica or Matlab or R, or advanced GP features like ensembles like FlexGP, etc. We chose to use a package that was more mature on the advanced features, but less mature in the user experience aspects. This choice is suitable for users with a high degree of expertise, but as we will see later, has its downside for both novice users and integration with other systems and prototyping. The process used to create a competitive GP solution for our Data Science task was as follows:

1. Feature selection as in GBR,
2. Simplification of mathematical operators,
3. Increased the training data size. Initial results showed that GP benefited with more data.

4. Leveraged the same sliding window approach for training the model. Initial results showed, that like GBR, GP extrapolation capability greatly reduced as newer data was received.
5. Added a validation set to first select for better solutions to extrapolate, but then later to enable an ensemble approach.
6. Added ensembles using a simple averaging of best solutions from the ensemble.

The biggest implication of these steps was that the code had to be changed in the GP system to output solutions to create the ensemble method, which means that a prototype system became much more brittle as a custom library as well as more custom scripts would need to be maintained.

Regarding the infrastructure required to run the GP system as a competitive Data Science tool, we leveraged our access to a very large cluster of compute nodes with multiple processors, each with a large amount of dedicated memory. This allowed us to develop GP solutions in a somewhat reasonable timescale as it took the GBR method. The GBR could develop models within several seconds on a basic desktop machine, approximately 17 s on average. For GP, one iteration took approximately 2 to complete one run. We executed this 30 times in parallel. Without parallelization this would take an hour. Additional parallelization could bring this down to around 5 min. While the compute time puts the GP method at a distinct disadvantage, in the era of Big Data, this kind of infrastructure difference is less critical: data sets will become larger and larger, giving easily parallelizable and distributed methods like GP an advantage, and data sets will natively be stored in massively distributed storage systems.

We now describe some of the GP settings we used. First we set the population size to 1000, and used a generation based model with 100 max generations. We used standard functions (+, −, *, /, sqrt, square, exp) and 32 variables. Other parameters were set to typical competent GP values: tree initial depth was 15, tournament selection was 10, crossover rate of 0.7, 10 tries to produce a unique tree in crossover, mutation rate or 0.2, and replication rate of 0.1. After some brief, initial probing of hyper parameters in the setup, we determined the following as an effective GP approach. First we split out training data into five equal sized time sequential groups. The first three earliest in time periods were used to train our GP system. To train, we ran our GP system 30 times to produce a selection of best solutions of various sizes and accuracy. Secondly, we used the fourth training period (the next consecutive one) as a validation data to select top 15 models (measured according to RMSE and model complexity) across all the populations and runs. Specifically, for each run we store the models that had the best accuracy and were the least complex (smallest), the least complex model, and the most accurate models. So, after 30 runs, we have approximately 90 possible models. From those models, we test them all against the fourth training period and score them by their RMSE. Then, we select the top 15 models (we determined 15 by minimal trial and error) and allow them to extrapolate on each new data point (sensors $s_1 \ldots s_{N-1}$) to predict sensor s_N. The final prediction is then the average of the predictions from the 15 models, as well as a 1 standard deviation of confidence bounds using all the predictions. There were approximately 240 lines of scripts created on top of the GP system to create the approach.

3.6.1 Visualization

While it is not common to visualize GBR solutions directly, the GP solutions can be viewed to understand variable relationships and overall system structure. Because we used an ensemble approach, it was a little less clear what or how the final solutions was calculated, but the clients appreciated seeing very clearly how attributes were used in the ensemble. To accomplish this, because the GP system we used did not allow easy viewing of the solutions, we created a Matlab script that converted the parenthesized in-fix notation from the GP system to a more easy readable one for the user. The matlab code utilizes Matlab symbolic math toolbox to perform the following actions:

- Replace variable names in equation
- Latex format the strings
- Display the strings as graphs

Our particular GP system works only with the variable name formats such as X1, X2, ... XN. An example model generated from the training is:

$$(((exp(((((X8 + X27)/(X32)^2) * sqrt(X30)) - X10))) * ((X1 + X4))^2).$$

This makes the equations hard to understand. The first operation performed using Matlab symbolic toolbox is to replace the variables to their proper names by specifying the correct symbol names for each variable. This is done as a batch process for the entire output population at the same time. The resulting output looks something like this:

$$((exp(((((MotorLoad + TemperatureC)/(TemperatureH)^2 * sqrt(TemperatureF)) - MotorTemperature))) + ((MotorSpeed + FlowRate))^2).$$

In the next step, the built-in Matlab latex() command converts a symbolic expression string into a latex formatted string as shown below:

$$e^{\frac{\sqrt{TemperatureF}\,(MotorLoad+TemperatureC)}{TemperatureH^2}-MotorTemperature} + (FlowRate + MotorSpeed)^2$$

The equation can then be displayed by creating text using Matlab Latex interpreter and then drawing the figure. The Matlab code to do this is very simple and consists of 33 lines of code.

3.6.2 Diversity

Diversity measures the variation between the population members. In evolutionary search, it is typically good to have the population be as diverse as possible while still improving accuracy. Diversity can be analyzed by several methods, for example:

- Variation between the existing terms in programs across the population. If many different programs have recurring terms, then there is less diversity.

- Spread of accuracy values for the population. If all programs in the population are giving similar accuracy, then they are all very similar. However, if the accuracies range is large, then the population is more diverse.

During typical GP model building, the accuracy stops improving at some point. As the accuracy improves, the diversity tends to decline before the accuracy has stopped improving. This is because as the accuracy is increasing, the more fit member of populations are generating offspring that are similar to them. Eventually the population becomes less diverse because all the members of the population are descended from similar fit individuals. So even though the population as a whole becomes more accurate, it becomes less diverse. In Data Sciences applications, diversity is a potential indicator of when to stop a run early, which could enable GP to have faster iteration time. In addition, having diverse sets of models is useful for creating ensembles of models. After the GP run loses diversity, the models are likely to have similar errors, or residuals, which when combined with each other in an ensemble is likely to be less beneficial. Ensemble methods benefit when solutions are combined that have different errors. Our initial experiments showed promise to use diversity as an early stopping criterion as well as a way to find better ensembles.

4 Lessons Learned

In the previous section, we described how two methods were used to solve the operations optimization problem for a real-world Data Science client. We also discussed several advantages, disadvantages and considerations of each method. Several lessons were extracted from our experiences that could be useful for future Data Science engagements as well as for future development of GP as a Data Science tool.

1. Data Science needs data management, and GP needs better linkage to the 'data environment'. Without a strong linkage to data management, GP must rely on additional tools to prepare data, intermediate data files, extra scripts to manage those intermediate files, which all create an additional burden on the Data Scientist and possibly the application developer who will turn the model into a prototype. GP systems like DEAP and Data Modeler are examples where a linkage to the data environment is strong.
2. The development of more competent GP systems needs to continue to be pursued, making the out-of-box performance of GP better. System enhancements during model development by a Data Scientist should not occur within core libraries. Elements like ensembles should be made as default.
3. Simplification of understanding and visualization of solutions can differentiate GP in the Data Science category. While some tools are better than others, the community should embrace this feature with new and insightful ways to optimize and visualize solutions—the effort should be very low to query a population of solutions for "how" they are solving the problem. Current approaches of listing

un-optimized, or un-simplified, postfix expressions, or counting frequent subtrees or variables, is not effective and could be potentially misleading.

4. GP needs to be able to iterate faster and reduce the time to create a good first model. Diversity could be an indicator of when to stop the model building process sooner and return a result. Other code optimization opportunities to reduce compute time should be pursued.

5. Frequent updating a model or adding new features during model re-training can confuse the end user and make maintenance difficult. When a new feature is added by the GP algorithm as a result of retraining the model with new data, the change should be intuitive to the user. A new direction of research could look to illuminate what new features used during model retraining might mean for solving or modeling the system.

6. Open Source implementations that have been matured by communities of users, as is the case for our GBR approach, are typically high performing out-of-the-box. When this is the case, it is best not to tweak the runtime parameters that have been optimized by others over long periods of time. GP should seek such broad community development to improve method robustness and out-of-the-box performance.

7. Code optimization should be saved for much later in the development phase. Giving priority to 'working scripts' leads to quick results, which can be shared with clients and approaches can be altered based on user feedback. Practitioners should avoid modifying the core GP system in favor of tuning system parameters contained in configuration files.

In our work, by comparing the two Data Science approaches, GBR and GP, we were able to see quite clearly the strengths and weaknesses of GP as compared to GBR. The above lessons learned represent both the positive attributes of GP as well as places where more work is needed. We see a lot of potential in GP as a new Data Science tool, particularly for use on Big Data and in complex, nonlinear, and domain knowledge intense domains. However, to get mainstream adoption, we believe these lessons learned should help identify future areas of both research and development of GP systems.

5 Conclusions

This chapter described a case study of applying GP to a real-world Data Science task in the problem domain of operations optimization. We believe the application is quite novel in attempting to build an online sensor estimation method to both validate data quality (it could be used to signal when a sensor is starting to drift), provide an estimation of a sensor when it fails or goes offline, as well as provide transparency to dynamic systems when they change by highlighting how the underlying GP solution changes. While both methods were able to find acceptable and similar accuracy, the GBR method won out in the client application for two reasons. Firstly, the code base of the GBR method consisted of open source code that could be easily implemented in a prototype environment by application engineers. The GP system contained too much unsupported code and many scripts that it required for data management, ensemble learning, diversity measurement, and visualization. Secondly, the training process was far simpler (both in number of lines and amount of time) for the GBR system. The GP system required more time to retrain and many intermediate files had to be managed. This chapter also introduced Data Science and emphasized the value of having a Data Management solution tightly coupled with GP. We leveraged an Ontology-Based Data Access approach using Semantic Web technologies. We hope that these contributions inspire future work that enable GP to become an effective Data Science tool.

References

Arnaldo I, Veeramachaneni K, O'Reilly UM (2014) Flash: A GP-GPU ensemble learning system for handling large datasets. In: Nicolau M, et al. (eds.) 17th European conference on genetic programming. LNCS, vol. 8599. Springer, Granada, pp 13–24

Castillo F, Kordon A, Sweeney J, Zirk W (2004) Using genetic programming in industrial statistical model building. In: O'Reilly UM, Yu T, Riolo RL, Worzel B (eds.) Genetic programming theory and practice II, Chap. 3. Springer, Ann Arbor, pp 31–48

De Rainville FM, Fortin FA, Gardner MA, Parizeau M, Gagne C (2012) DEAP: a python framework for evolutionary algorithms. In: Wagner S, Affenzeller M (eds.) GECCO 2012 evolutionary computation software systems (EvoSoft). ACM, Philadelphia, PA, pp 85–92

Dhar V (2013) Data science and prediction. Commun ACM 56(12):64–73

Dubcakova R (2011) Eureqa: software review. Genet. Program. Evolvable Mach. 12(2):173–178

Fazenda P, McDermott J, O'Reilly UM (2012) A library to run evolutionary algorithms in the cloud using MapReduce. In: Di Chio C, et al. (eds.) Applications of evolutionary computing, EvoApplications 2012, LNCS, vol. 7248. Springer, Malaga, pp 416–425

Friedman J (2001) Greedy function approximation: a gradient boosting machine. Ann Stat 29(5):1189–1232

Gustafson S, Sheth A (2014) Web of things. Computing Now 7(3). http://www.computer.org/web/computingnow/archive/march2014

Icke I, Bongard J (2013) Improving genetic programming based symbolic regression using deterministic machine learning. In: de la Fraga LG (ed.) 2013 IEEE conference on evolutionary computation, Cancun, vol. 1, pp 1763–1770

Jones E, Oliphant E, Peterson P, et al. (2001) Scipy: open source scientific tools for python. http://wwwscipyorg

Kordon AK, Smits GF (2001) Soft sensor development using genetic programming. In: Spector L, et al. (eds.) Proceedings of the genetic and evolutionary computation conference (GECCO-2001). Morgan Kaufmann, San Francisco, CA, pp 1346–1351

Koza JR (1992) The genetic programming paradigm: Genetically breeding populations of computer programs to solve problems. In: Soucek B, the IRIS Group (eds.) Dynamic, genetic, and chaotic programming. Wiley, New York, pp 203–321

Loukides M (2010) What is Data science? OReilly Radar Report. http://cdn.oreilly.com/radar/2010/06/What_is_Data_Science.pdf

O'Neill M, Vanneschi L, Gustafson S, Banzhaf W (2010) Open issues in genetic programming. Genet Program Evolvable Mach 11(3/4):339–363 (tenth Anniversary Issue: Progress in Genetic Programming and Evolvable Machines)

Pedregosa F, Varoquaux G, Gramfort A, Michel V, Thirion B, Grisel O, Blondel M, Prettenhofer P, Weiss R, Dubourg V, Vanderplas J, Passos A, Cournapeau D, Brucher M, Perrot M, Duchesnay E (2011) Scikit-learn: machine learning in python. J Mach Learn Res 12:2825–2830

Schmidt M, Lipson H (2009) Distilling free-form natural laws from experimental data. Science 324(5923):81–85. doi:10.1126/science.1165893. http://ccsl.mae.cornell.edu/sites/default/files/Science09_Schmidt.pdf

Smits GF, Vladislavleva E, Kotanchek ME (2010) Scalable symbolic regression by continuous evolution with very small populations. In: Riolo R, McConaghy T, Vladislavleva E (eds.) Genetic programming theory and practice VIII. Genetic and evolutionary computation, Chap. 9, vol. 8. Springer, Ann Arbor, pp 147–160

van der Walt S, Colbert SC, Varoquaux G (2011) The numpy array: a structure for efficient numerical computation. Comput Sci Eng 13:22–30

van Harmelen F, Hendler JA, Hitzler P, Janowicz K (2015) Semantics for big data. AI Magazine 36(1):3–4

Veeramachaneni K, Arnaldo I, Derby O, O'Reilly UM (2015) FlexGP: Cloud-based ensemble learning with genetic programming for large regression problems. J Grid Comput 13(3):391–407

Wagner S, Kronberger G (2011) Algorithm and experiment design with heuristiclab: an open source optimization environment for research and education. In: Whitley D (ed.) GECCO 2011 tutorials. ACM, Dublin, pp 1411–1438

The Evolution of Everything (EvE) and Genetic Programming

W.P. Worzel

Abstract The Internet is entering a new period of growth driven by an increasing number of processors connected at the edge of the Internet. Many of these processors are sensors that continuously collect data. By 2020, it is projected that there may be more than 20 billion (1000 million) devices connected to the Internet. Collectively these devices are called the Internet of Things (IoT) or the Internet of Everything (IoE). The sheer volume of the data that will be gathered creates new problems for an economy that is increasingly driven by data analytics. It is likely that the devices at the edge of the Internet will take part in the processing of data for analytics by using distributed computing among edge devices. Genetic Programming could play a unique role in this environment because of its ability not only to gather and analyze data, but to control the evolution and use of other machine learning algorithms. The confluence of unimaginable streams of real-world data and emergent behaviors may give rise to the question of whether the evolution of intelligence in the natural world can be recreated using evolutionary tools.

Keywords Internet • Internet of Things • Fog Lifter • Combinators • SKGP • P2P • AllJoyn • IOx • FRP • Blockchain • Functional relational programming • Data flow design • Evolutionary reinforcement learning

1 Background

Currently data is collected from many devices across the Internet and uploaded to data centers where the data is processed in the aggregate. The notion of the Internet of Things (IoT) is characterized by an increasing number of devices at the edge of the Internet gathering data for focused purposes. Many of these devices are sensors or packages of sensors. Specialized industrial sensors connected to the Internet are used for many purposes such as chemical production facilities, the power grid, and agriculture. Homes and specialized personal devices such as personal fitness and health devices, entertainment centers, and thermostats are also connected to the

W.P. Worzel (✉)
Fog Lifter Inc., 1314 Beechwood Dr., Ann Arbor, MI 48103, USA
e-mail: billwzel@gmail.com

© Springer International Publishing Switzerland 2016
R. Riolo et al. (eds.), *Genetic Programming Theory and Practice XIII*,
Genetic and Evolutionary Computation, DOI 10.1007/978-3-319-34223-8_8

Internet. City infrastructure and buildings have an abundance of sensors as well, including water sensors, traffic sensors, and environmental sensors. As the number and variety of devices increases, it creates an increasing computational burden to filter and integrate data in real time. The estimated level of 20 billion new devices will make the model of sending raw data for processing unsustainable both in terms of the communications infrastructure and the raw processing power needed at data centers. In particular, finding, collecting and analyzing data in real time from this many devices becomes an impossible task. Instead it will be necessary for some level of data processing to be done at the edge and the results will be integrated locally and shared globally. Cloud Computing assumes that bulk of computing done will in 'The Cloud' which usually means at distant data servers such as those provided by Amazon, Google and other providers of computing resources. Most uses of the Internet of Things rely on The Cloud to process data generated at the edge. In contrast to Cloud Computing, the term 'Fog Computing' is used to describe a more diffuse form of computing that is closer to the ground and where analytics is done locally and potentially shared with many users. This approach is already becoming an important part of the discussion. Cisco has introduced an operating system called IOx in Cisco (2014a) that is already being used in industrial applications, including in Smart Grid applications described in Cisco (2014b). The AllSeen Alliance (2012) has created an open source OS called AllJoyn that is particularly useful for in-home devices. Meanwhile Google has recently announced a version of its Android OS called KitKat described in Google (2015) that is optimized for low memory devices such as is common in Edge Computing. But while these advances are important, they focus only on the local aspects of Fog Computing rather than a vision that integrates the Fog with the Cloud; it is assumed that the Cloud as it exists currently is all that is necessary to take advantage of the coming data explosion in the Fog.

2   Fog Lifter™

The author has been working collaboratively on a suite of open source software that will facilitate the full integration of Fog Computing into the Internet. Fog Lifter is built on the assumption that local computation and results will be used for multiple purposes, allowing geographically dispersed information to be combined with other data for very different purposes. This may be summarized as "Compute locally, analyze anywhere" where the results of local computation is made available through a registry and may be used as part of a larger computation, whether in the Cloud or at another 'locality'.

2.1 *Example Uses of Fog Lifter*

Some examples of the sort of use cases where Fog Lifter would be advantageous are shown in Figs. 1 and 2. In Fig. 1, data from agricultural fields are collected from sensors placed in or near fields and gathered into a model of each field which, when collected together, produces a model for the entire farm. The data from these models may be shared with other local farmers, who may be part of the same Farm Coop, to give a local view of the crops being grown and the state of the soil and water use in the area. These can then be aggregated regionally to inform local markets, crop insurance agents and farm machine usage planning. Note that often this is not raw data but processed results from the raw data collected from the fields.

Figure 2 shows how this approach can be extended to a global scale if the data is made available to other interested parties, up to, and including global entities. Here local or regional data is not only used on larger geographical areas, but selected portions of the analysis done at a local area used for different purposes.

For example, governmental water planners may not be interested in regional crop production, but are very interested in agricultural water usage as part of larger forecast of water needs that include industrial and civilian water use. This information may also be integrated with weather forecasts that influence potential agricultural water usage as well as the availability and state of water sources such as aquifers, reservoirs and snow packs.

However, multinational NGOs may be interested not only in water usage and long range weather models, but agricultural yields and the spread of pests in order to predict where there are likely to be crop shortages and danger of starvation.

While this kind of planning already exists, by combining data from thousand or millions of sensor sources related to farming, weather, insect control, and water source and usage, the quality and accuracy of data and predictions could be greatly improved.

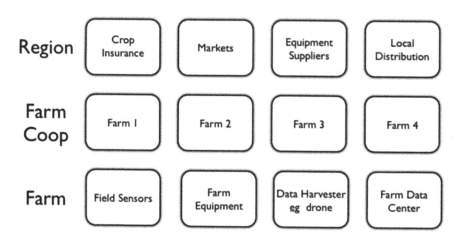

Fig. 1 Aggregation of local analytics

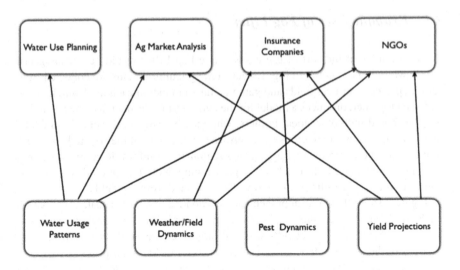

Fig. 2 Aggregated data used for global models

In these scenarios, what does Fog Computing look like? Sensors may be buried or even plowed into fields and the data can be collected during the growing season whenever farm machinery or drones pass over the field to collect the data. Beyond simple collection of data, each sensor, while not connected to a central data repository such as the farm center, may be connected wirelessly to other sensors. Since modern sensors have a great deal of processing capacity, but potentially limited storage capacity (particularly when data is collected 24-7), by collaboratively processing data using their combined processing capabilities, they can "compress" the data into a more concise form. In this context, one may consider data processing as the ultimate compression algorithm.

Similar examples may be found in weather prediction, traffic models, city infrastructure, the Smart Grid and many manufacturing industries.

2.2 Fog Lifter Platform

In order for Fog Lifter to accomplish this goal, it must meet the following criteria:

1. Organize distributed computing of local data;
2. Support intermittent connectivity;
3. Make locally processed results available globally;
4. Integrate local results with the Cloud (remote data centers) or other fog localities (local clusters of computational resources).

To accomplish this, Fog Lifter has the following design components:

1. Functional Relational Programming using combinators as functional "microcode";
2. A data registry;
3. Conceptual data flow design;
4. Security and Privacy protections.

Each of these will be described briefly.

2.2.1 Functional Relational Programming (FRP)

Moseley and Marks (2006) stated that with the growing complexity of software, it was necessary to produce more robust software. Toward this end they suggested an approach they called functional relational programming that used functional programming techniques to avoid unintentional side effects, particularly when code is expected to be distributed between processors. Relational programming structures are combined with functional programming to embody the idea of immutable data to avoid unintentional side effects being embedded in the data that is stored from one result and passed to another program later. They also use logic programming techniques to constrain possible values of data (e.g., specifying that a value can never be allowed to reach 0).

This idea is already in use for large scale analytics in data centers and has been shown to produce fewer errors. This is a large part of why languages such as Scala as described in Wampler (2014) and Clojure, described in Rochester (2013) and libraries such as Cascading, described in Nathan (2013) are growing in popularity. The adoption of this approach to Fog Computing is therefore natural as it is imperative that programming distributed across hundreds or even thousands of devices minimize the chance for failure.

To implement FRP, Fog Lifter uses combinators as a way to abstract local variables based on Turner (1979). This allows any pure functional code to be compiled into code that can be run on any device and components of the code can be distributed safely among processors. By applying combinator expressions to relational data structures, and enforcing strict, temporally limited storage of data during computation, Fog Lifter adheres to the most important aspects of FRP.

2.2.2 Data Registry

The function of the data registry is to provide both a semantic description of data that is published on the Internet as well as publishing a data dictionary of the format of the data. This dictionary takes the form of a relational data dictionary to facilitate the use of FRP in Fog Lifter.

2.2.3   Conceptual Data Flow Design

Data Flow Design is usually used as a documentation tool that explicates the flow of data and processing within a system, often within a database. However, it can be used to describe the flow of data in a distributed application as it describes how and when data is distributed, what processing is done in parallel and how the data is joined after processing.

An example of a Conceptual Data Flow Design diagram is shown in Fig. 1. Here the flow of data is shown by the "pipes" connecting computational actions and the tagged nodes show the processing of relations. Each of the blue 'M' labels denote a map process and the red 'R' labels denote a reduce operation. While this is a graphical diagram, the same information can be rendered in a table format in a concise form that is easy to process. In Fog Lifter these design elements can be connected from sources through processors to sinks where the results are stored. Implicit in this is that a result from one analysis may be made available through the Registry for other analyses (Fig. 3).

2.2.4   Security and Privacy

Security and privacy issues are already a major topic in the emerging world of the IoT/ IoE. A key assumption in Fog Lifter is that sharing of data and computing resources must be an act of permission, not one of omission. To implement this, a transactional approach via blockchain technology is used. The use of data, or processors must either be certified by a transaction permitting the use to a specific entity or paid for by that entity and the owners control whether they are willing to share with any given user. Details about such approaches may be found in Wood (2015) and Brody and Pureswaran (2015).

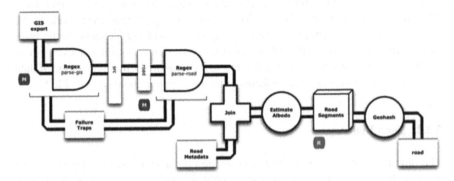

Fig. 3 Data workflow design (from Nathan 2013)

3   The Evolution of Everything (EvE)

What if we combined the locality and real-world data flow made available by the Internet of Things with the evolutionary power of genetic programming? How would we design such a system? What would be the result?

Since GP is usually used with a limited set of inputs and a very focused purpose as represented in a fitness function, what happens when the inputs are expanded to include a significant part of the real world and the purpose is broadened to a more general and less single-valued focus? The IoT makes it possible to conceive of such a situation and to expand the application of GP to more open-ended problems.

Paradoxically, it is the added constraints of the real world that increase the usefulness of GP. These include the constraints of physics, resource allocations and timeliness which may be difficult to capture in an artificial environment but are intrinsically present in a real-world setting.

3.1   Non-trivial Geography

In Spector and Klein (2005), the authors describe a GP system where they use a very simple notion of locality only for the purpose of crossover. While many GP systems use an amorphous breeding population where any individual can combine with any other, Spector and Klein added a simplistic, linear structure to the geography of the population to constrain the potential breeding population to a limited number of individuals within a finite distance from an individual as potential breeding partners. This simple change produced solutions that were much more effective in outcome (i.e., better fitness) and more economical (i.e., less computational effort to produce) than results produced without a notion of geography.

While the notion of geography was not new to evolutionary algorithms since, as Spector and Klein noted, others had used complex simulation environments with distinct geographies or sub-populations (demes) that were complex in representation and difficult to manage, Spector and Klein showed that the use of even a simple geography seemed to lead to significant improvement at minimal cost. In essence, they had developed a reductio ad absurdum test to the notion of individual locality and the result had been a startling increase in performance. Their tentative conclusion from a GP perspective was that limiting the mating pool increased diversity among locales. They also point out that because of the simplicity of implementation, it could easily be adopted in systems that already have a notion of geography (such as deme based systems) or even in systems where sub-populations are defined by performance, such as ALPS (Hornby 2006) or other performance segregated sub-populations.

In the IoT, many devices have a distinct location in the real world. For example, farms, water sources, power sources and cars all attach great significance to location. By tying GP selection for crossover to real-world locations, GP would find localized solutions and, when combined with other local solutions, regional solutions emerge. Conversely, some solutions would be so dependent on a very limited location, that solutions that were not highly localized would fail. This suggests that a meta-analysis of geography to solution effectiveness would help "redraw the map" of a field of study. For example, identifying microclimates could significantly change weather predictions for an area.

3.2 Evolutionary Reinforcement Learning (ERL)

Ackley and Littman (1991) describe a system that combines neural nets and a genetic algorithm in a simulated world. The idea was to evolve creatures that could learn from their environment using neural nets to recognize things in the environment and then act based on the signals produced. To accomplish this, there are two sets of neural nets: those that respond to sensory input and those that take action in response to the inputs. The weights in the sensory neural net are heritable and fixed. The weights used in the action neural net are refined using a reinforcement learning algorithm during the life of the individual where weights are reinforced if the environment is evaluated as being better at time t+1 than it was at time t.

The genetic algorithm is used to create successive generations of creatures, mostly by crossover of encoded values that are mapped to weights for the neural nets. Successful individuals in the environment propagate based on the life or death of the individuals living in the artificial world. Details of the artificial world and the behaviors engendered by this system are not described here, but the reader is encouraged to read the details of the dynamics of the system and the population of individuals in the above cited paper as they are informative and interesting in their relationship to population dynamics as described in biological literature.

However, the principles of ERL lend themselves to the IoT as well. Here neural nets would be used to process data from a sensor, and actions would be taken based on the inputs from the sensors. The results of the actions taken across groups of sensors would dictate the evolution of sensor behavior. An example might be weather prediction where the accuracy of prediction based on present data and past adaptation could lead to more precise forecasts.

3.3 The SKGP

Worzel and MacLean (2015) described the use of combinators as described in Turner (1979) as the basis of a GP system that used a Hindley-Milner type system (Hindley 1997). As detailed in Worzel and MacLean (2015), it has many virtues, including first class functions, and natural partitioning of functional components that work well in a map-reduce environment, and (unique to GP) the fact that

Fig. 4 Particulate genes in GP

components tend to be reused across generations means that the results of an application to a set of data may be cached for reuse during fitness evaluations. It also avoids some of the intrinsic limits that are associated with tree-based GP as described in Daida (2003).

In the context of the IoT, there is an additional extension to the SKGP that is needed for EvE: the addition of particulate genes. Freeland (2003) strongly advocated for an adoption of particulate genes in GP. By 'particulate genes' what is meant in this context is the notion that sections of code and/or values are exchanged as a whole with a section of code (a particulate gene) taken from one parent or the other intact. Using the SKGP, particulate genes could be developed where each gene consisted of a combinator function that could be applied to a specific set of data and, potentially, be applied to one another. Figure 4 shows an example of a particulate set of genes.

Moreover, one could envision a meta-function applied to a list of particulate genes in such a way that the meta-function would select and apply some of the individual gene-expressions to one another based on values produced because of the context of the values produced in each gene. In this way particulate genes could mimic the removal of introns during the post-transcription modification of RNA.

Figure 6 shows a function being applied to a list of expressions and data in order to select and combine elements of the particulate gene, effectively combining the elements in the list according to the function being applied to it.

From an evolutionary perspective, particulate genes have many advantages. Crossover can now be particulate, with a complete "gene" from either of two parents being selected and possibly mutated, leaving the structure of each gene intact and avoiding the disruption of tree-based crossover. It is also easier to flag individuals as belonging to a particular species, or gender and allows dominant/recessive genes, all of which improve population dynamics.

3.4   GP Reinforcement Learning (GPRL)

If the SKGP is combined with the ideas behind ERL, while using particulate genes, then neural networks can be embodied in a single gene and crossover can proceed in a manner similar to the way it is done in ERL with parent weights being combined during crossover. Essentially Evaluation neural nets and Action neural nets become genes in a particulate gene approach to GP.

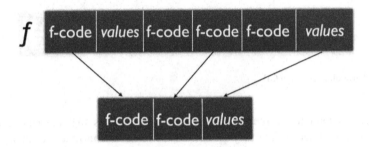

Fig. 5 Modification of particulate genes from meta-function f

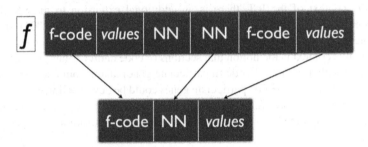

Fig. 6 GPRL with particulate genes

Similarly a function that selects and combines genes based on the values returned could be implemented. In this scenario, the example in Fig. 5 is extended to the design shown in Fig. 6.

3.5 Assembling EvE

Putting all these elements together by using Fog Lifter to provide a sense of place for the functions in a GP population, adapting ERL to work within GP using an SKGP-like system and a meta-function that selects and combines particulate genes provides the basis for a continuous learning system. For example, over time, an agricultural system could continuously refine crop management recommendations as it moves from passively learning to actively recommending actions, or a city infrastructure could adapt to new construction, changing economics, or even macro changes such as climate-change related issues.

4 Discussion

The Genetic Programming Theory and Practice workshop was designed to encourage speculative ideas. Parts of EvE are highly speculative. Fog Lifter is a work in progress, but much of the code is drawn from open source software that is quite robust. The SKGP exists and has been applied to commercial and research problems, particularly in the biotech space. However, it does not yet use particulate genes, instead relying on directed graphs to replace trees of standard "Koza Style" GP. While it is safe to say that these graphs can be contained within the conceptual notion of a particulate gene, this has not yet been implemented.

Similarly, while ERL as described in Ackley and Littman (1991) has been implemented, the author has not had access to the code base. It is probably a good idea to revisit ERL in this context.

The adaptation that uses particulate GP with specific genes encompassing the output of neural nets as part of a "post-transcriptional" removal of "exons." is (as far as the author has been able to determine) a wholly new idea in the GP world though there has been some use of the word 'exon' to denote "code bloat" in the GP literature (see for example Soule 2002). Here, the meta-function may, in one situation, select one set of genes for use in the overall function and in another circumstance, select others. In this, it must be confessed, the concept is closer to epigenetic phenomena or splice variants in biology than simply removing exons.

Beyond the mechanics of how EvE might work, there is the underlying idea that moving from simulated worlds to real-time, real-world data, creates a sea change for evolutionary algorithms. Instead of being limited by imprecise models of a world that evolves for a set number of generations, it is based on continuous, adaptive evolution. While it is not expected that such a system would instantly create a high-precision model of the real world, over time it is expected that valuable predictive models will emerge. Moreover, the failures of such systems can teach us much about the world. The author has long contended that we learn as much or more from analyzing the failures of GP to make correct predictions as we do from successes as described in. Assumptions about the real world are often mistaken and GP often makes these mistakes clear during runs that incorporate these assumptions.

Beyond simply modeling the world, (Holland 1995, 1998), talk at length about the property of emergence of new properties and behaviors and describes what he believes is the key components of such systems: State, Transition Function, Generators and Agents. EvE maps fairly well onto these elements as follows:

1. State: the internal state of an individual function in response to the current environment;
2. Transition Function: the application of an evolved function in response to the environment;
3. Generators: the real world (no need to simulate an environment in the IoT!);
4. Agents: individuals in the environment.

Following Holland, this suggests that at a minimum, such a system could evolve new and unexpected behaviors as it interacts with the environment by prescribing actions in the real world (e.g., when to plant or harvest a field or what route a car should take). Taken to the limit, one could speculate that a complex system could "come alive"—providing, of course, that there is a consensus of what being alive means! In the realm of pure speculation then, as the IoT grows, and our capacity to distribute processing of information across devices, one may begin to approach the question of whether, and how, the Singularity could occur. Perhaps the best way to approach this is to point out that we have one example of evolution reaching intelligence in the natural world on earth. If the singularity is reachable in-silico (a proposition about which the author is skeptical), then our best plan of would be by mimicking the mechanism of evolution while using inputs from the natural world.

At this point we reach the end of the known world and the author will end by repeating an excerpt of a poem by Alice Fulton used in Holland (1995):

> ... Its point of view? One
> with the twister in vista glide,
> and the cricket in the ditch,
> with riverrain, and turbines' trace.
> Inside the flux of
> flesh and trunk and cloudy come,
> within the latent
> marrow of the egg, the amber
> traveling waves is where
> its vantage lies.
> Entering the tornado's core,
> entering the cricket waltzed by storm–
> to confiscate the shifting give
> and represent the with-
> out which.

References

Ackley D, Littman M (1991) Interactions between learning and evolution. In: Langton C, Taylor C, Farmer C, Rasmussen S (eds) Artificial life II. SFI studies in the science of complexity, vol X. http://www2.hawaii.edu/nreed/ics606/papers/Ackley91learningEvolution.pdf

AllJoyn (2012) Documentation. Tech. rep., AllSeen Alliancex, https://allseenalliance.org/developers/learn

Almal AA, Mitra AP, Datar RH, Lenehan PH, Fry DW, Cote RJ, Worzel WP (2006) Using genetic programming to classify node positive patients in bladder cancer. In: Keijzer M, Cattolico M, Arnold D, Babovic V, Blum C, Bosman P, Butz VB, Coello C, Dasgupta D, Ficici SG, Foster J, Hernandez-Aguirre A, Hornby G, Lipson H, McMinn P, Moore J, Raidl G, Rothlauf F, Ryan C, Thierens D (eds.) GECCO 2006: Proceedings of the 8th annual conference on genetic and evolutionary computation. ACM, New York, pp 239–246

Brody P, Pureswaran V (2015) The next digital gold rush: How the internet of things will create liquid, transparent markets

Cisco (2014a) Cisco fog computing with iox. Tech. rep., Cisco. "http://www.cisco.com/c/dam/en/us/products/collateral/se/internet-of-things/at-a-glance-c45-732380.pdf"

Cisco (2014b) Utilities/smart grid. Tech. rep., Cisco. http://www.cisco.com/c/en/us/solutions/industries/energy/external-utilities-smart-grid.html

Daida JM (2003) What makes a problem GP-hard? A look at how structure affects content. In: Riolo RL, Worzel B (eds) Genetic programming theory and practice, Chap 7. Kluwer, Boston, pp 99–118. doi:10.1007/978-1-4419-8983-3_7. http://www.springer.com/computer/ai/book/978-1-4020-7581-0

Freeland S (2003) Three fundamentals of the biological genetic algorithm. In: Riolo RL, Worzel B (eds) Genetic programming theory and practice Chap 19. Kluwer, Boston, pp 303–311. doi:10.1007/978-1-4419-8983-3_19. http://www.springer.com/computer/ai/book/978-1-4020-7581-0

Google (2015) Android kitkat. https://developer.android.com/about/versions/kitkat.html

Hindley J (1997) Basic simple type theory. Cambridge University Press, Cambridge

Holland J (1995) Hidden order: how adaptation builds complexity. Addison-Wesley, Redwood City

Holland J (1998) Emergence: From chaos to order. Addison-Wesley, Reading

Hornby (2006) ALPS: the age-layered population structure for reducing the problem of premature convergence. In: GECCO 2006: Proceedings of the 8th annual conference on Genetic and evolutionary computation, vol 1. ACM, pp 815–822

Moseley B, Marks P (2006) Out of the tar pit. "http://web.mac.com/benmoseley/frp/paper-v101.pdf"

Nathan P (2013) Enterprise data workflows with cascading. O'Reilly Media, Sebastopol

Rochester E (2013) Clojure data analysis cookbook. Packt Publishing, Birmingham

Soule T (2002) Exons and code growth in genetic programming genetic programming. In: Proceedings of the 5th European conference, EuroGP. LNCS, vol 2278. Springer, pp 142–151

Spector L, Klein J (2005) Trivial geography in genetic programming. In: Yu T, Riolo RL, Worzel B (eds) Genetic programming theory and practice III, genetic programming, vol 9, Chap 8. Springer, Ann Arbor, pp 109–123. doi:10.1007/0-387-28111-8_8. http://hampshire.edu/lspector/pubs/trivial-geography-toappear.pdf

Turner D (1979) A new implementation technique for applicative languages. Softw Pract Exp 9:31–49

Wampler D (2014) Why scala is taking over the big data worldwhy scala is taking over the big data world. http://www.slideshare.net/deanwampler/why-scala-is-taking-over-the-big-data-world

Worzel WP, MacLean D (2015) SKGP: The Way of the Combinator. In: Riolo R, Worzel WP, Kotanchek M (eds) Genetic programming theory and practice XII. Genetic and evolutionary computation. Springer, Ann Arbor, pp 53–71.

Wood G (2015) Yellow paper: Ethereum's formal specification. Tech. rep., Ethereum. https://github.com/ethereum/yellowpaper

Lexicase Selection for Program Synthesis: A Diversity Analysis

Thomas Helmuth, Nicholas Freitag McPhee, and Lee Spector

Abstract Lexicase selection is a selection method for evolutionary computation in which individuals are selected by filtering the population according to performance on test cases, considered in random order. When used as the parent selection method in genetic programming, lexicase selection has been shown to provide significant improvements in problem-solving power. In this chapter we investigate the reasons for the success of lexicase selection, focusing on measures of population diversity. We present data from eight program synthesis problems and compare lexicase selection to tournament selection and selection based on implicit fitness sharing. We conclude that lexicase selection does indeed produce more diverse populations, which helps to explain the utility of lexicase selection for program synthesis.

Keywords Lexicase selection • Diversity • Tournament selection • Implicit fitness sharing

1 Introduction

Lexicase selection is a recently developed parent selection method for evolutionary computation in which individuals are selected by filtering the population according to performance on individual fitness cases, considered in random order (Spector 2012). Lexicase selection, when used as the parent selection method in genetic programming, has been shown to provide significant improvements in terms of problem-solving power (Helmuth et al. 2014; Helmuth and Spector 2015). In this chapter we investigate the reasons for the success of lexicase selection, focusing in

T. Helmuth (✉)
Computer Science, University of Massachusetts, Amherst, MA, USA
e-mail: helmutht@wlu.edu

N.F. McPhee
Division of Science and Mathematics, University of Minnesota, Morris, MN, USA
e-mail: mcphee@morris.umn.edu

L. Spector
Cognitive Science, Hampshire College, Amherst, MA, USA
e-mail: lspector@hampshire.edu

© Springer International Publishing Switzerland 2016
R. Riolo et al. (eds.), *Genetic Programming Theory and Practice XIII*,
Genetic and Evolutionary Computation, DOI 10.1007/978-3-319-34223-8_9

particular on the ways in which lexicase selection seems to help maintain population diversity. We present data from eight program synthesis problems and compare lexicase selection, in terms of problem solving power and diversity, to tournament selection and selection based on implicit fitness sharing (IFS). IFS distributes reward among the individuals that solve a test case, giving more reward for cases solved by fewer individuals (McKay 2000); for more detail see Helmuth et al. (2014).

For each parent selection event lexicase selection randomly orders the test cases and then removes any individuals that do not have the best performance on the first case. If more than one individual remains then those that do not have the best performance on the second case are also removed. This continues until only one individual remains and is selected, or until all cases have been used, in which case one of the remaining individuals is selected randomly. Key properties of lexicase selection are that (a) it avoids combining all errors into a single scalar fitness value, (b) because of the random ordering of test cases, every test case will be most important (first to be considered) at least occasionally, and (c) similarly, each pair of test cases, and each triple, etc., will be most important at least occasionally.

We investigate the relations between selection methods and population diversity using two measures of diversity: error diversity and cluster counts. We find that lexicase selection runs have consistently higher error diversity than tournament selection and IFS across all generations and all problems. The cluster counts for lexicase selection are also generally higher, but less consistently. We conclude that lexicase selection does indeed produce more diverse populations, which helps to explain the utility of lexicase selection for program synthesis.

2 Diversity Measures

To evaluate a program in program synthesis, we run it on a set of test cases composed of input/output pairs, creating a behavior vector of its outputs. Then, we apply one or more error functions to each desired output and the program's output, creating an error vector for each individual. We define *error diversity* to be the percentage of distinct error vectors in the population. Error diversity is similar to *behavioral diversity*, which is the percentage of distinct behavior vectors in the population (Jackson 2010). The error diversity of a population will be less than or equal to its behavioral diversity, since two different behavior vectors may produce the same error vector, but two different error vectors must come from different behavior vectors. Helmuth et al. (2014) showed that lexicase selection maintained higher diversity than tournament and IFS selection on three problems.

One hypothesis we have put forth regarding the improved performance of lexicase selection is that it enables groups of specialists for solving different parts of the problem to evolve side-by-side, implicitly maintaining the kind of niches that are maintained more explicitly by island models and related methods. We expect that evolution may sometimes progress when individuals from different groups mate, producing a child that combines the abilities of its parents. The hope is that this process, iterated, will eventually produce an individual that solves the entire problem. Here we explore the effects of different parent selection methods on the development of clusters of individuals that perform similarly across the test cases. We expect that using lexicase selection will result in relatively larger numbers of clusters, since it selects individuals on the basis of specific cases and groups of cases, rather than on overall performance.

To examine this idea, we must be able to measure the clustering of a population with respect to the training cases. We base the clustering of the population on the individuals' error vectors across the training cases. Since we are primarily interested in whether an individual performs at least as well as every other individual in the population, we convert the error vectors into binary "elitized" error vectors that indicate whether an individual achieved the best error on each test case in that generation. More formally, if each individual j in the population P has error vector $error_j$ containing error values on the test cases T, then the elitized error vector for individual i is defined by

$$elitized_i[t] = \begin{cases} 0, & \text{if } error_i[t] = \min_{j \in P}(error_j[t]) \\ 1, & \text{otherwise} \end{cases}$$

for $t \in T$. By elitizing the error vectors, we can ignore the differences between individuals that perform poorly on cases in different ways, and concentrate on how individuals cluster based on the cases on which they perform well.

In this work we use agglomerative clustering[1] to count how many clusters there are in the population at each generation. Agglomerative clustering creates a hierarchical clustering model by first placing each individual into its own cluster. It then iteratively combines the two closest clusters into a single cluster, until all clusters have been combined into a single cluster, recording at each step the distance between the clusters in each merged pair. We can then break the single cluster into smaller clusters by "cutting" the merge between any two clusters whose distance exceeds some threshold. Since we are using binary error vectors, we use the Manhattan distance as our distance metric, which makes the distance between two error vectors a count of the number of test cases on which those two individuals have different "eliteness" results. We chose to count the number of clusters that differed on at least 10% of the training cases; for example, if a problem has 200 training cases, we count the number of clusters that differ in binary eliteness on at least 20 training cases. While this distance is somewhat arbitrary, it gives a reasonable and consistent estimate of how many groups of individuals are doing significantly different things in a given generation.

---

[1]We used the agnes (Maechler et al. 2014) implementation of agglomerative clustering in R (R Core Team 2014), using the average linkage when combining clusters.

3   Experiment and Results

We collected data from 100 runs each on eight different problems described by
Helmuth and Spector (2015). All of these are basic programming problems taken
from introductory programming texts; several are readily solved, while others
remain unsolved using this study's tools. Table 1 lists the problems, a brief
description, and the length of the error vectors[2]; other details of the runs can be
found in Helmuth and Spector (2015). In Table 2 we've also provided the number of
successes, i.e., runs in which a program was evolved with total error of 0 across
all the training cases. Success rates aren't the focus of this chapter, but these
numbers give a sense of the relative difficulty of the problems and illustrate the
substantial improvements that lexicase selection provides over both tournament
selection and IFS.

Table 1 Short descriptions of the eight test problems used here, along with the number of errors
in each error vector

Problem name	Description	# errors
Replace space with newline	Print the input string, replacing spaces with newlines. Also, return the number of non-whitespace characters	200
Syllables	Count the number of occurrences of vowels (a, e, i, o, u, y) in the given string and print that number as X in The number of syllables is X	200
String lengths backwards	Given a vector of strings, print the length of each string in reverse order (starting with last and ending with first)	100
Negative to zero	Given a vector of integers, return the vector where all negative integers have been replaced by 0	200
Double letters	Given a string, print the string, doubling every letter character, and tripling every exclamation point. All other non-alphabetic and non-exclamation characters should be printed a single time each	100
Scrabble score	Given a string of visible ASCII characters, return the Scrabble score for that string	200
Checksum	Given a string, compute the integer ASCII values of the characters in the string, sum them, take the sum modulo 64, add the integer value of the space character, and then convert that integer back into its corresponding character (the checksum character). Then print Check sum is X, where X is replaced by the correct checksum character	200
Count odds	Return the number of odd numbers in a vector of integers	200

See Helmuth and Spector (2015) for more details on each problem

---

[2]For some of these problems, each test case generates multiple error values because we apply more
than one error function.

Table 2 Number of successes (out of 100 runs) for each of the eight test problems used here. These numbers are similar but not identical to those reported in Helmuth and Spector (2015) because new runs were performed for this chapter

Problem name	Lexicase	Tournament	IFS
Replace space with newline	57	13	17
Syllables	24	1	2
String lengths backwards	75	18	12
Negative to zero	72	15	9
Double letters	5	0	0
Scrabble score	0	0	0
Checksum	0	0	0
Count odds	4	0	0

We used the Clojush implementation[3] of the PushGP system (Spector and Robinson 2002; Spector et al. 2005) for all runs. Each run used a population size of 1,000 individuals, and runs continued for either 300 generations or a until solution was found, whichever came first.

Figures 1, 2, 3, 4, 5, 6, 7, 8, 9, 10, 11, 12, 13, 14, 15, and 16 show error diversity and cluster counts over time for each of the test problems. Below each plot is a smaller sub-plot showing the number of successes over time for each selection; since runs end when a solution is found, the successes plot gives a sense of how many runs are still being represented in the primary plot at a given generation. In Fig. 1, for example, the number of lexicase successes is nearly 25 by generation 50, and nearly 50 by generation 150. Thus there are slightly more than 75 data points still represented in the lexicase data at generation 50, but only about 50 data points represented from generations 150 to 300. Each plot includes a line indicating the median error diversity or median cluster count across whichever of the 100 runs was still running at that generation. We also indicate the range from the 25th percentile to the 75th percentile with a gray band around the median line; unfortunately the tournament and IFS results are often very similar and strongly overlap, making them difficult to differentiate.

In general the error diversity numbers for lexicase selection are substantially and significantly higher than those for either tournament selection or IFS, which tend to be extremely similar. The String Lengths Backwards problem was the only problem for which there was any substantial overlap between the range of values for lexicase and the other two selection mechanisms (see Fig. 5). Typically the lexicase error diversity rises very sharply in the early generations leveling off somewhere between 0.75 and 1.0, meaning that $\frac{3}{4}$ or more of the individuals in the lexicase runs have unique error vectors. This is in contrast to the tournament selection and IFS results, in which the median error diversity values rarely rise above 0.5; the two exceptions are on the Scrabble Score and Count Odds problems (Figs. 11 and 15), which neither ever solved, where the error diversity values approach or exceed 0.75.

The cluster count results are more mixed. Lexicase selection has clearly higher cluster counts for half of the problems (Replace Space With Newline, Syllables,

---

[3]https://github.com/lspector/Clojush

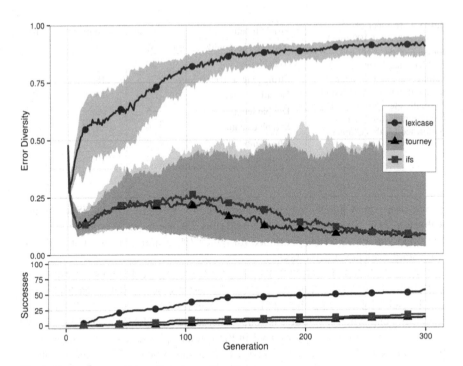

Fig. 1 Replace space with newline—error diversity

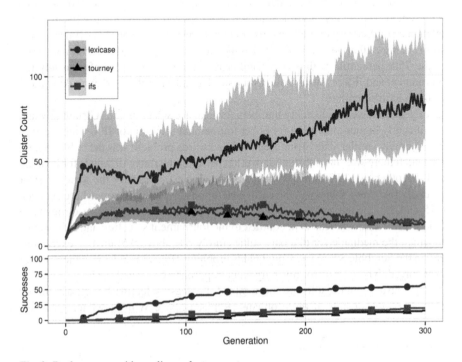

Fig. 2 Replace space with newline—cluster counts

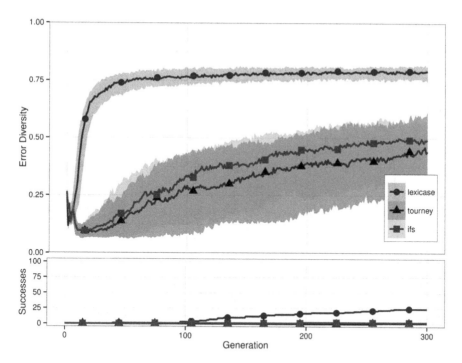

Fig. 3 Syllables—error diversity

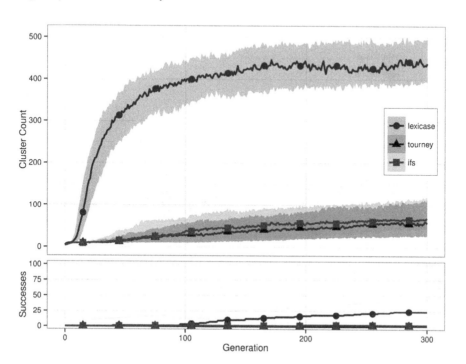

Fig. 4 Syllables—cluster counts

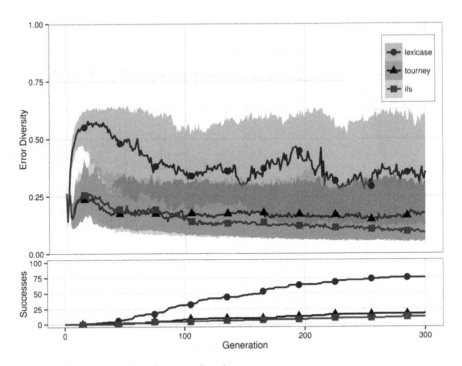

Fig. 5 String lengths backwards—error diversity

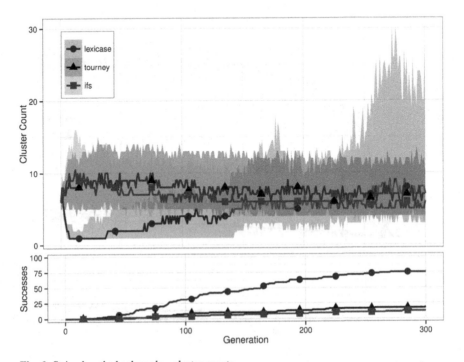

Fig. 6 String lengths backwards—cluster counts

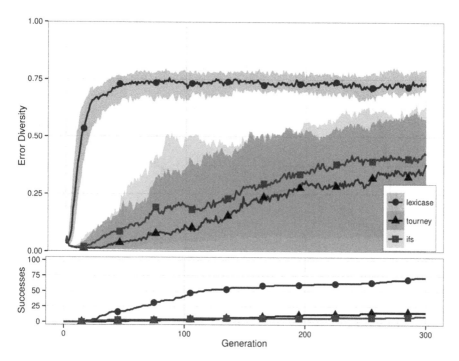

Fig. 7 Negative to zero—error diversity

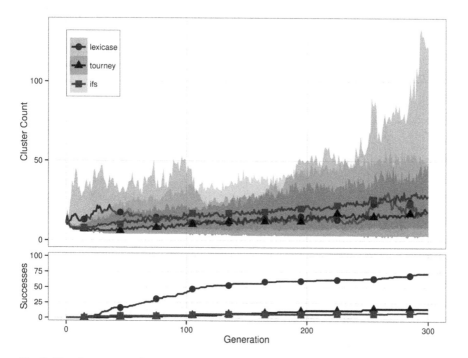

Fig. 8 Negative to zero—cluster counts

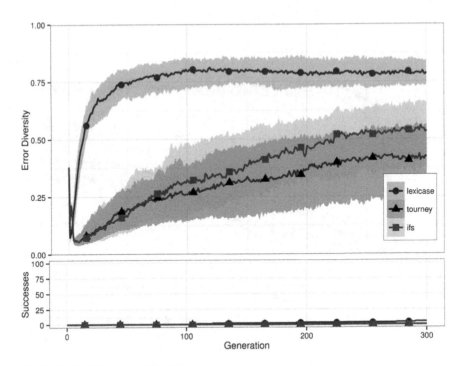

Fig. 9 Double letters—error diversity

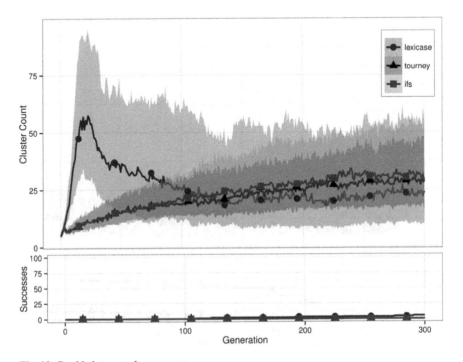

Fig. 10 Double letters—cluster counts

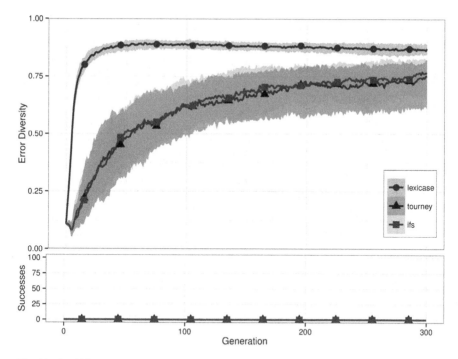

Fig. 11 Scrabble score—error diversity

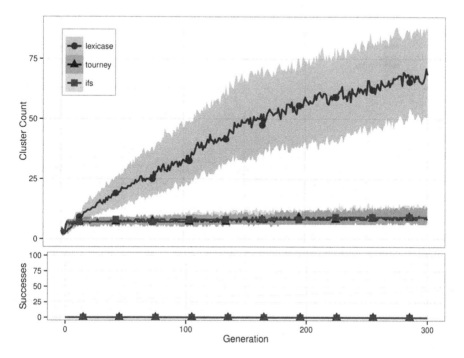

Fig. 12 Scrabble score—cluster counts

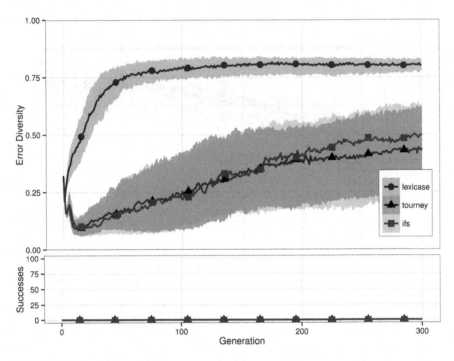

Fig. 13 Checksum—error diversity

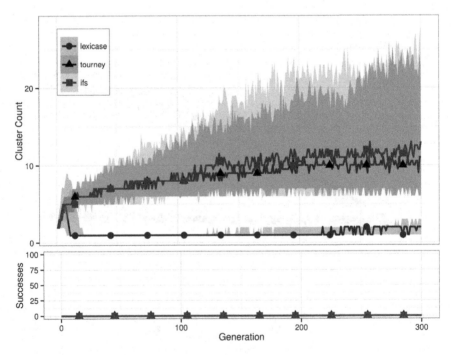

Fig. 14 Checksum—cluster counts

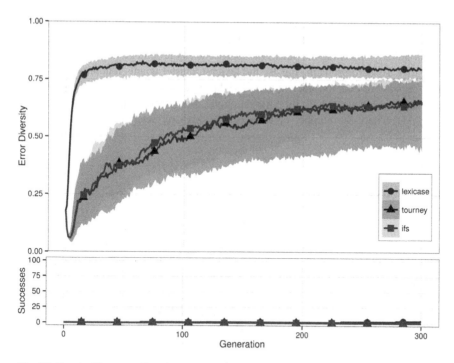

Fig. 15 Count odds—error diversity

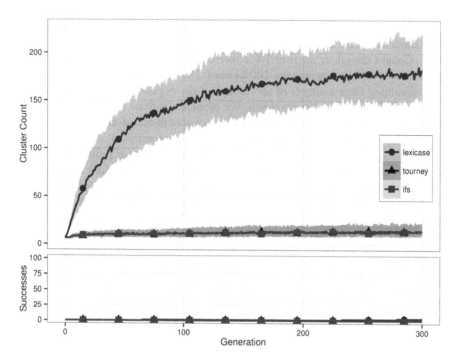

Fig. 16 Count odds—cluster counts

Scrabble Score, and Count Odds; Figs. 2, 4, 12 and 16). It also starts with much higher counts on the Double Letters problem (Fig. 10), but those numbers drop again quickly, matching the other two approaches by around generation 100. On the Negative To Zero problem (Fig. 8), the lexicase cluster counts remain small (about the same as for both tournament and IFS) throughout the runs. Particularly striking are lexicase cluster counts for String Lengths Backwards (Fig. 6) and Checksum (Fig. 14), where the number of clusters with lexicase selection is actually lower earlier in the run.

4 Discussion

As in Helmuth and Spector (2015), lexicase selection produced more successes than either tournament selection or IFS on any problem in which a solution was found. The error diversity for the lexicase runs was much higher than for tournament and IFS for most problems, which is consistent with the hypothesis that lexicase selection helps maintain diversity. The lexicase error diversity values tended to plateau at or above 0.75, meaning that in a population of 1000 individuals there were over 750 *distinct* error vectors. This doesn't mean that different individuals were *solving* different test cases; it could just be that many had different incorrect answers and error values. From a search perspective, though, this still seems useful, as those different error values may represent different starting points for subsequent search.

For four of the eight problems, the cluster counts were also much higher for lexicase than for the other two selection mechanisms. For some of these problems (e.g., Count Odds) there are over 100 clusters, and for Syllables the median cluster count is over 400 from generation 100 forward. This suggests that lexicase selection is maintaining large numbers of sub-groups of the population that are capable of solving different parts of the problem. For problems with no solutions found, this might indicate that the genetic operators are not able to act on the structure of the programs in those sub-populations in ways that allow progress.

Interpretation of the cluster count results on the other four problems is more difficult. Analysis of the lexicase Checksum runs suggests that the lack of clustering might be a function of structural issues with the test cases; there are 100 test cases, with two error functions per test case: the Levenshtein edit distance on the printed string, and the integer difference between the ASCII values of the last character of the printed string and the correct checksum character. It appears that populations quickly evolve the ability to print Check sum is, but then stall, with each program printing different final characters. This allows for fairly high error diversity (over 0.75), but any given program tends to get at most two or three test cases right by guessing. This means that the Manhattan distance between any two elitized error vectors is typically only 5 or 6 at most, shy of the 10 % threshold of 20 for this problem, resulting in only one or two clusters. Additional test cases exploring different inputs might allow evolution to first stumble upon and then exploit code that produces actual checksums.

On problems for which solutions were discovered, lexicase selection runs found solutions throughout the 300 generations. This, combined with the high levels of error diversity and the often high number of clusters, gives one hope that meaningful search can still occur late in a lexicase selection run. The plots of successes over time under the primary plots typically appear to have positive slope even at generation 300, so it would be interesting to extend these runs to 500 or 1000 generations and see how many additional solutions are discovered. If lexicase selection is indeed maintaining meaningful diversity then we would expect to see continued discovery of solutions, at a higher rate than for either tournament selection or IFS. This might be particularly interesting for problems for which solution discovery is rare but possible, such as Double Letters and Count Odds, which are solved using lexicase selection 5 and 3 times respectively, but not at all using tournament selection or IFS. Solutions for these two problems tended to be discovered later in the run (Double Letters in generations 109, 122, 192, 275, and 291; Count Odds in 65, 233, 279), so letting runs on those problems go longer might be revealing.

On the set of problems explored here, error diversity seems to be a better predictor of performance than cluster counts. In fact, on two of the problems for which solutions were found in over half the runs (String Lengths Backwards and Negative To Zero), lexicase selection maintained very small numbers of clusters, similar to tournament and IFS. On the other hand, lexicase selection consistently maintained higher error diversity than other methods, and found more solutions on every problem that was solved. This may indicate that the ability to form clusters on a problem is more indicative of the problem itself than the parent selection method and its ability to solve the problem. This provides evidence against our hypothesis that lexicase selection performs better because it maintains clusters of individuals that genetic operators can combine to solve increasingly large numbers of test cases.

5 Conclusions

In this chapter we used two different measures of diversity (error diversity and cluster counts) to try to better understand the impact of lexicase selection, and why it seems to consistently outperform tournament selection and implicit fitness sharing (IFS) on a range of software synthesis problems (Helmuth and Spector 2015). The error diversity was generally *much* higher for lexicase selection than for either tournament selection or IFS, with lexicase selection maintaining a broad range of distinct behaviors. Cluster counts were typically higher with lexicase selection, and the instances in which they weren't may say more about the problem or test case structure than about the selection mechanism. This suggests that error diversity is indeed a valuable metric for studying the impact of system design decisions. The value of cluster counts is less clear, but it seems likely that understanding why the cluster counts were so low on certain problems could be informative.

Given that the lexicase selection runs maintain error diversity all across the 300 generations, it seems plausible that extending the length of the runs would generate additional solutions. It would be illuminating to extend these runs to 500 or 1000 generations and see whether lexicase selection is able to make "better" use of those additional computational resources.

While the focus of this chapter was to better understand the behavior of lexicase selection, the results also show that tournament selection and IFS behave *very* similarly with respect to the diversity measures used here. This is unfortunate because IFS was specifically designed to maintain diversity. Both tournament selection and IFS aggregate test case errors into a single value, with IFS just weighting the components differently; this may be partially responsible for the similar rates in diversity.

Acknowledgements Thanks to the members of the Hampshire College Computational Intelligence Lab for discussions that helped to improve the work described in this chapter, to Josiah Erikson for systems support, and to Hampshire College for support for the Hampshire College Institute for Computational Intelligence. This material is based upon work supported by the National Science Foundation under Grants No. 1017817, 1129139, and 1331283. Any opinions, findings, and conclusions or recommendations expressed in this publication are those of the authors and do not necessarily reflect the views of the National Science Foundation.

References

Helmuth T, Spector L (2015) General program synthesis benchmark suite. In: Silva S, Esparcia-Alcazar AI, Lopez-Ibanez M, Mostaghim S, Timmis J, Zarges C, Correia L, Soule T, Giacobini M, Urbanowicz R, Akimoto Y, Glasmachers T, Fernandez de Vega F, Hoover A, Larranaga P, Soto M, Cotta C, Pereira FB, Handl J, Koutnik J, Gaspar-Cunha A, Trautmann H, Mouret JB, Risi S, Costa E, Schuetze O, Krawiec K, Moraglio A, Miller JF, Widera P, Cagnoni S, Merelo J, Hart E, Trujillo L, Kessentini M, Ochoa G, Chicano F, Doerr C (eds) GECCO '15: Proceedings of the 2015 on genetic and evolutionary computation conference. ACM, Madrid, pp 1039–1046. doi:10.1145/2739480.2754769. http://doi.acm.org/10.1145/2739480.2754769

Helmuth T, Spector L, Matheson J (2014) Solving uncompromising problems with lexicase selection. IEEE Trans Evol Comput. doi:10.1109/TEVC.2014.2362729

Jackson D (2010) Promoting phenotypic diversity in genetic programming. In: Schaefer R, Cotta C, Kolodziej J, Rudolph G (eds) PPSN 2010 11th international conference on parallel problem solving from nature. Lecture notes in computer science, vol 6239. Springer, Krakow, pp 72–481. doi:10.1007/978-3-642-15871-1_48

Maechler M, Rousseeuw P, Struyf A, Hubert M, Hornik K (2014) Cluster: Cluster Analysis Basics and Extensions. R package version 1.15.3

McKay RI (2000) Fitness sharing in genetic programming. In: Proceedings of the genetic and evolutionary computation conference. Morgan Kaufmann, Las Vegas, pp 435–442

R Core Team (2014) R: A language and environment for statistical computing. R Foundation for Statistical Computing, Vienna. http://www.R-project.org/

Spector L (2012) Assessment of problem modality by differential performance of lexicase selection in genetic programming: a preliminary report. In: 1st workshop on understanding problems (GECCO-UP). ACM, Philadelphia, pp 401–408. doi:10.1145/2330784.2330846

Spector L, Robinson A (2002) Genetic programming and autoconstructive evolution with the push programming language. Genet Program Evolvable Mach 3(1):7–40. doi:10.1023/A:1014538503543

Spector L, Klein J, Keijzer M (2005) The push3 execution stack and the evolution of control. In: GECCO 2005: Proceedings of the 2005 conference on Genetic and evolutionary computation, vol 2. ACM, Washington, pp 1689–1696. doi:10.1145/1068009.1068292

Behavioral Program Synthesis:
Insights and Prospects

Krzysztof Krawiec, Jerry Swan, and Una-May O'Reilly

Abstract Genetic programming (GP) is a stochastic, iterative generate-and-test approach to synthesizing programs from tests, i.e. examples of the desired input-output mapping. The number of passed tests, or the total error in continuous domains, is a natural objective measure of a program's performance and a common yardstick when experimentally comparing algorithms. In GP, it is also by default used to *guide* the evolutionary search process. An assumption that an objective function should also be an efficient 'search driver' is common for all metaheuristics, such as the evolutionary algorithms which GP is a member of. Programs are complex combinatorial structures that exhibit even more complex input-output behavior, and in this chapter we discuss why this complexity cannot be effectively reflected by a single scalar objective. In consequence, GP algorithms are systemically 'underinformed' about the characteristics of programs they operate on, and pay for this with unsatisfactory performance and limited scalability. This chapter advocates *behavioral program synthesis*, where programs are characterized by informative execution traces that enable multifaceted evaluation and substantially change the roles of components in an evolutionary infrastructure. We provide a unified perspective on past work in this area, discuss the consequences of the behavioral viewpoint, outlining the future avenues for program synthesis and the wider application areas that lie beyond.

Keywords Genetic programming • Program behavior • Program semantics • Multiobjective evaluation • Search driver • Evaluation bottleneck

K. Krawiec (✉)
Computational Intelligence Group, Institute of Computing Science, Poznan University of Technology, Poznań, Poland
e-mail: krawiec@cs.put.poznan.pl

J. Swan
Department of Computer Science, University of York, York, UK

York Centre for Complex Systems Analysis, University of York, York, UK

U.-M. O'Reilly
ALFA, Computer Science and Artificial Intelligence Laboratory (CSAIL), Massachusetts Institute of Technology (MIT), Cambridge, MA, USA

© Springer International Publishing Switzerland 2016
R. Riolo et al. (eds.), *Genetic Programming Theory and Practice XIII*,
Genetic and Evolutionary Computation, DOI 10.1007/978-3-319-34223-8_10

1 Introduction and Motivations

Program synthesis is a challenging task due to the size of the search space, its multimodality, externalized semantics of instructions, and complex contextual interactions between them. These characteristics are intrinsic to the nature of this task and cannot be evaded. However, some difficulties faced by contemporary genetic programming (GP), in particular the far from satisfactory scalability, result from the particular model of evaluation of candidate solutions adopted in this generative, trial-and-error metaheuristic.

As in the majority of genres of evolutionary computation (EC), the candidate solutions (programs) in GP are conventionally evaluated using scalar, generic performance measures. Such a measure will usually capture program error, e.g. represented either as the number of failed tests (for discrete domains) or the total error committed on them (for the continuous domains).

The practice of measuring the quality of candidate programs using a scalar performance measure has several merits. It allows for strict and elegant formulation of a program synthesis task as an optimization problem, and is thus compatible with the conventional way of posing problems in artificial intelligence, operational research, and machine learning. It also eases the separation of generic search algorithms from a domain-specific evaluation function, which is so vital for *meta*heuristics. No wonder that this 'design pattern' is so common that we rarely ponder its other consequences.

Unfortunately, there is a price to pay for all these conveniences, which arises from the inevitable loss of information that accompanies the process of scalar evaluation. That loss is particularly high in generate-and-test program synthesis like GP, where not only a program itself is a complex combinatorial entity, but also its execution is an intricate iterative process. In consequence, the spectrum of possible *behaviors* exhibited by programs is enormously rich. For example, even when looking only at program output, the number of all possible behaviors of programs that attempt to solve the (trivial for contemporary GP) problem of 6-bit multiplexer is the staggering 2^{64}. Yet, in conventional GP all that is left of that process is a single number (in the interval $[0, 64]$ for the above example). The conventional scalar evaluation *denies a search algorithm access to the more detailed information on program's behavioral characteristics*, while *that information could help to drive the search process more efficiently*.

This observation can be alternatively phrased using the message-passing metaphor typical in information theory. A search algorithm and an evaluation function can be likened to two parties that exchange messages. The message the algorithm sends to the evaluation function encodes the candidate solution to be evaluated. In response, the algorithm receives a return message—the evaluation. In a sense, the evaluation function *compresses* a candidate solution into its evaluation. If one insists on compressing all the information about program behavior into a scalar fitness that aggregates various aspects of that behavior, then one also has to accept the fact that such compression is inevitably lossy.

This *evaluation bottleneck* has detrimental consequences. The outcomes on particular tests compensate each other and may render programs indistinguishable in a selection phase, leading to loss of diversity and premature convergence. Also, tests may vary with respect to objective difficulty (the probability of a random program passing a test), subjective difficulty (measured by search algorithm's likelihood to find a program that passes the test), or both. In consequence, evolution often tends to greedily synthesize programs that pass the easiest tests, and such programs may correspond to local minima in the search space. These and other properties of conventional evaluation cause it to exhibit low fitness-distance correlation (Tomassini et al. 2005), i.e. to not reflect well the number of search steps required to reach the optimal solution. As a result, guiding search by a fitness function defined in this way may be not particularly efficient. In other words, the fitness function, despite embodying the objective quality of candidate solutions (considered as prospective outcomes of program synthesis process), is not necessarily the best driver to guide the search.

The parsimony of conventional evaluation is also awkward in architectural terms, i.e. when looking at a program synthesis system as a network of interconnected components. Why would one component (fitness function) compress the evaluation outcomes and then force another component (search algorithm) to reverse-engineer them, knowing that this incurs loss of information? There are no reasons for this other than the convention inherited from metaheuristic optimization algorithms and evolutionary metaphor.

Arguably, there are domains where an evaluation function is by definition 'opaque' and makes this bottleneck inevitable. For instance, in Black Box Optimization, fitness is the only information on a candidate solution available to a search algorithm. However, it might be the case that the need of such separation is more an exception than a rule when considering the whole gamut of problems we tackle with metaheuristics. In many domains, there are no principal reasons to conceal the details of evaluation, which is often complex and an abundant source of potentially useful information. This is particularly true for program synthesis, where the act of evaluating a candidate program is rich at least in two respects. Firstly, a program interacts with *multiple tests*, and will often perform differently on each. Secondly, a program's confrontation with a single test involves executing *multiple instructions*.

The main motivation for this chapter is the observation that the habit of driving search using a conventional, scalar evaluation function cripples the performance of stochastic program synthesis as implemented by GP. In response, we posit the necessity of broadening the evaluation bottleneck and providing search algorithms with more detailed information on program behavior. This leads to a new paradigm of *behavioral program synthesis*. In this chapter, we demonstrate a particular means to this end, presented earlier in preliminary forms in Krawiec and Swan (2013) and Krawiec and O'Reilly (2014), which relies on the concept of information-rich *search drivers*, alternative quasi-objectives that may be capable of guiding program synthesis process more efficiently than the conventional objective function.

2 Behavioral Program Synthesis

In this section we sketch the vision of *behavioral program synthesis*, a methodology for program synthesis that prioritizes program behavior.

Several existing extensions of the traditional GP paradigm involve broadening of the evaluation bottleneck, in a more or less explicit way. For instance, program semantics in GP is the vector of program outputs for particular tests (Moraglio et al. 2012) and thus provides more information about program behavior than the conventional scalar fitness. Behavioral characterizations like program semantics are tailored to the needs of a particular approach: a semantics of a program holds program output for every test, because this is the information required by (most) semantic-aware search operators.

Contrary to this model, we propose that evaluation provides a complete account of program behavior, and to leave it up to the other components of a search algorithm to decide which pieces of that information to use. The means for this are *program trace*, which reports the detailed, instruction-by-instruction effects of program execution for a given input, and *execution record* that gathers and aligns such traces for all considered tests.

Both these concepts can be conveniently explained with an example. Figure 1 presents an integer-valued symbolic regression task ('Problem') defined by four tests, each of them comprising two input variables x_1 and x_2 and the desired output y. Assume the tree-based GP program p shown there ('GP Individual') needs to be evaluated. The colored lists present the outcomes of intermediate execution stages, produced by p at particular instructions for consecutive tests. When gathered together, they form the execution record (labeled 'ML dataset' in Fig. 1, for the reasons explained later). A single row in an execution record captures the behavior of p on the corresponding test in the set of tests; for instance, the first row does so for $x_1 = 2$ and $x_2 = 3$. For this input, the intermediate values generated by p at consecutive instructions are $2, 2, 3, 2, 4, 5$, when executing p in the bottom-top, left-to-right manner (the ordering could be different for this side effect-free programming language). The corresponding first row of the execution record presents this in an abridged form, where the duplicates are omitted: $2, 3, 4, 5$ (the second and the fourth leaf in the tree refer to the input variable x_1 that has been already recorded in the first element of the trace).

A trace is thus a sequence of intermediate computation states, and can be harvested from a running program by interrupting its execution after every instruction, and taking a 'snapshot' of the *execution environment*. In the above example with functional tree-based GP, a state is simply the working value returned by the currently executed node of an expression tree. Other representations used in GP require different implementations of this concept. In linear GP (Brameier and Banzhaf 2007), statements operate via side-effects i.e. by changing the values stored in registers; the environment there would be the states of all registers. In the PushGP system (Spector and Robinson 2002), where the working memory is the code

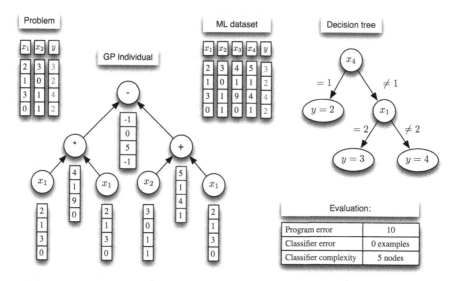

Fig. 1 The workflow of behavioral evaluation in Pattern-guided Program Synthesis (PANGEA, Krawiec and Swan 2013), valid also for behavioral programming presented in Krawiec and O'Reilly (2014)

stacks and data stacks, all these data structures taken together form the execution environment. Nevertheless, in both these cases recording traces is straightforward, as demonstrated by our use of PushGP in Krawiec and Swan (2013).

Differences between program representations notwithstanding, an execution record captures the entirety of effects of program's interactions with the input data provided in tests. As such, it is obviously possible to derive from it the conventional fitness (by comparing the last column with the vector of desired outputs), the outcomes of interactions with individual tests [which opens the door to posing a program synthesis task as a test-based problem (Popovici et al. 2011)], or program semantics [in the sense of semantic GP (Moraglio et al. 2012)]. The arguably most exciting possibility (which has been little explored to date) lies in investigating 'internal' program behavior, which we attempt in the following.

For expressions like the one in the above example, the execution record is by definition *aligned*, i.e. the states recorded in the same column correspond to the same instruction in the program. For programs containing loops, conditional statements, or involving recursion, traces may have different length and alignment is not guaranteed. Nevertheless, this does not invalidate our hypothesis that certain regularities present in an execution record can be valuable telltales of program's actual or prospective performance. The particular approach presented in the next section exemplifies this claim.

3 Pattern-Guided Program Synthesis

In conventional GP (and other conceivable generate-and-test program synthesis techniques), candidate programs are normally judged by their outputs. However in GP, arguably more than in many other domains, the ultimate program output is an effect of collective effort of constituent instructions. One reason for this state of affairs is the sequential nature of programs. The other is the particularly complex mapping from program code to behavior: a minute modification of the former may cause a dramatic change in the latter. On the other hand, a major change in a program can turn out to be behaviorally neutral, due to the multimodality mentioned above.

It is thus likely that programs emerge in an evolving population that feature potentially useful components (subprograms, code fragments) yet that usefulness is not leveraged by the final instructions. Such programs will usually perform poorly in terms of conventional fitness and likely get lost in selection phase. Conventional GP has no means to counteract that loss. However, traces and execution records introduced in the previous section may reveal such intermediate *behavioral patterns*. Given that, it seems tempting to look for them in order to identify the subprograms that 'relate' to the task in question. Programs that feature such subprograms could be then promoted, to allow the search operators to turn them into better-performing candidate solutions. For instance, a fortunate crossover may mate such a promising subprogram with a piece of code that together leads to optimal solution. This acquired knowledge could be alternatively used more explicitly, for instance by archiving the subprograms and then reusing them via search operators.

A skilled human programmer may discover behavioral patterns and exploit them to design a program that meets the specification of a program synthesis task. Humans in general are known to be incredibly good at spotting and thinking in patterns when solving all sorts of problems—for this reason they have been termed *informavores* (Miller 1983). A sizeable part of AI research is about mimicking such capabilities (Hofstadter 1979). Moreover, humans can *anticipate* the patterns that are desirable in a given problem and often use domain and commonsense knowledge for that purpose. Consider the task of synthesizing a program that calculates the median of a list of numbers. The background knowledge tells us that a reasonable first stage of solving this task is to sort the list. In terms of execution records, reaching an intermediate execution state that contains the sorted elements of the list is desirable in this task.

To realize these opportunities, an efficient detector of 'interesting' (relevant for a given program synthesis task) behavioral patterns is necessary. One may for instance analyze how execution traces converge between tests, because this to some extent determines program output—if two or more traces arrive at the same execution state, their further courses must be the same, assuming that an execution state captures everything about the execution environment (by including, for instance, instruction pointer). In Krawiec and Solar-Lezama (2014), we proposed an approach that quantified program quality with respect to such convergences of traces, using concepts from information theory.

Nevertheless there exists a wider class of behavioral patterns of potentially greater interest, namely the patterns that are detectable by conventional knowledge discovery and machine learning (ML) algorithms. Such patterns are perused by the method described in the following, termed PANGEA (PAtterN Guided Evolutionary Algorithms), originally proposed in Krawiec and Swan (2013) and then extended in Krawiec and O'Reilly (2014). Technically, behavioral patterns are being revealed there by a ML algorithm trained on execution traces. Information on the resulting classifier is then used to augment the fitness function. By relying on generic ML tools, this process does not rely on domain knowledge (as is common for humans). Rather, it seeks abstract regularities that can be used to predict the correct output of a program. If this approach is able to reveal meaningful dependencies between partial outcomes and the desired output, we may hope to thereby promote programs with the potential to produce good results in future, even if at the moment of evaluation the output they produce is incorrect.

The ML perspective on behavioral program synthesis originates in the observation that an execution trace bears some similarity to an *example* in ML. Assuming the execution record resulting from applying p to all tests is aligned, i.e., the states in particular traces correspond to each other, the columns of the record can be likened to *attributes* in ML. The desired program output y corresponds in this context to the desired response of a classifier (or regressor, depending on the nature of the task). And crucially, a ML induction algorithm (*inducer* for short), given a set of such examples, can be used to produce a classifier that predicts the desired output of the program based on the attributes describing execution traces.

The method proceeds in the following steps, exemplified in Fig. 1:

1. An execution record is built by running the program on the tests.
2. The execution record is transformed into a conventional ML dataset D.
3. A ML induction algorithm is applied to D, resulting in a classifier C.
4. Program evaluation is calculated from the properties of C.

The record built in Step 1, as explained in the example in Sect. 2 (Fig. 1), is subsequently transformed in Step 2, resulting in the training set labeled as 'ML dataset' in the figure. In this case of simple tree-based GP, the attributes correspond one-to-one to the columns of the execution record, so the only essential change is the addition of the program output y as a dependent variable (class label) in the dataset. Transformation of an execution record into a ML dataset could be more sophisticated, for instance if states represent compound rather than elementary data types. More advanced transformation could facilitate discovery of behavioral patterns, for instance when representation biases of a ML classifier prevent it from capturing certain classes of pattern. Yet another motivation is to allow discovery of higher-order patterns that are unobservable when each attribute reflects a single execution state. Though these options deserve future research, here we focus on tree-based GP and the straightforward, one-to-one transformation of the columns of execution record into ML attributes.

Given the training set D, in Step 3 we train a ML classifier C on it. In Krawiec and Swan (2013) and Krawiec and O'Reilly (2014), we used the decision tree inducers

(C4.5 (Quinlan 1992) and REP-tree, respectively). For the example in Fig. 1, a decision tree induction algorithm produced the classifier labeled 'Decision tree', considering attributes x_i as well as the decision class y as nominal variables. The tree comprises five nodes, uses attributes x_4 and x_1 to predict the output of the program, and commits no errors on D.

3.1 Search Drivers

The classifier maps the attributes derived from intermediate execution states onto the desired output of the program. In a sense, it attempts to complement the program's capability for solving the problem (i.e. producing the desired output value). This observation motivates the design of specific evaluation functions. If the traces reveal regularities that are relevant for predicting the desired output, then the induction algorithm should be able to build a classifier that is (1) compact and (2) commits relatively few classification errors. These aspects are strongly related to each other, which we illustrate in the following.

Consider first the case of an optimal program p. p solves the task, i.e. produces the desired output $y_i = p(\mathbf{x}_i)$ for all tests $(\mathbf{x}_i, y_i) \in T$. Since each trace ends with a final execution state, and the attributes are collected from all states, then the last attribute in D will be among them. Because p solves the task, that attribute will be identical to the desired output. In such a case, the induction algorithm may produce a classifier of C that involves only that attribute, e.g. a decision tree composed of a single decision node and k leaves corresponding to the k unique desired outputs. Such a decision tree is thus quite compact and commits no classification errors.

Now consider a non-optimal program. Assume its output diverges so much from the desired output that the corresponding attribute is useless for prediction. In such a case, it is likely for the induced classifier to rely on the other attributes, derived from the intermediate execution states. Individually, such attributes have usually limited predictive power, unless the corresponding column in an execution record happens to capture some key aspect of the task. In consequence, the resulting classifier of C needs to rely on many such attributes and thus may be quite complex. In the case of decision trees, the tree will feature many decision nodes. In general, the size and predictive accuracy of the classifier depend on the degree to which the intermediate states *relate* to the desired output.

These examples illustrate that complexity and predictive capability of a classifier are related to each other in a nontrivial manner. Aggregating them would involve unnecessary loss of information, as we argued earlier. This motivated us to define *two* evaluation functions: the *classification error* and *classifier complexity*. The technical definition of the latter depends on classifier representation; for decision trees, it will be the number of tree nodes. Clearly, neither of these evaluation functions alone captures fully the relatedness of attributes to the desired output. It becomes natural to use them side-by-side. In Krawiec and Swan (2013), we aggregated them into a single evaluation function. In Krawiec and O'Reilly (2014), we kept them separate and relied on multiobjective approach, employing the Non-dominated Selection Genetic Algorithm (NSGA-II, Deb et al. 2002). NSGA-II relies on tournament selection on Pareto ranks to make the choices. To break the ties on ranks, it employs *sparsity*, a measure that rewards the candidate solutions that feature less common scores on criteria. The method is also elitist in selecting from the combined set of parents and offspring (rather than from parents alone).

We postulate that quantities like classifier error and classifier complexity (as well as the information-theoretical measures we proposed in Krawiec and Solar-Lezama 2014) share certain features in common and exemplify a new class of evaluation functions, which we refer to as *search drivers*. A search driver can be considered as a 'quasi' evaluation function. It is expected to provide a certain search gradient towards the global optima, but not necessarily a strong one—we posit that what matters is the *direction* of that gradient rather than its magnitude. We are particularly interested in search drivers that are uncorrelated with the original objective function, as this opens the possibility of using them (or multiple search drivers) together, preferably in a multiobjective setup. Also, we do not expect search drivers to be minimized at the optima—we find this requirement unnecessarily constraining when designing search drivers, while in GP program correctness can be easily verified in abstraction from evaluation function.

In EC, the concept that arguably most resembles that of search driver is *surrogate fitness*. Also known as *approximate fitness function* or *response surface* (Jin et al. 2002), a surrogate function provides a computationally cheaper approximation of the original objective function that comes with a given problem. Search drivers diverge however from surrogate fitness in several respects. Firstly, surrogate functions are by definition meant to *approximate* the original objective function. Search drivers lack this intent. Given the challenges plaguing conventional objective functions (see Introduction), why would one want to approximate them? Secondly, search drivers are intended to aid GP meant as a *search*, not optimization problem. This leaves more freedom in their design, which do not have to 'mimic' the objective function across the entire search space. Thirdly, in a program synthesis task, a search driver is not required to be consistent with the objective function in attaining minimal values at global optima. In surrogate fitness, such consistency is essential. And last but not least, a primary rationale for surrogate fitness is high computational cost of the objective function, while the role of search drivers is to help navigate more effectively in the search space.

These differences justify the conceptual distinctness of search drivers. In an ongoing work, we hope to provide a more sound formalization of this concept and come up with guidelines for principled design of search drivers.

3.2 *Experimental Evidence*

In Krawiec and Swan (2013) and Krawiec and O'Reilly (2014) we applied PANGEA to PushGP (Spector and Robinson 2002) and tree-based GP respectively (Fig. 2). In both cases, the behavioral approach systematically outperformed the

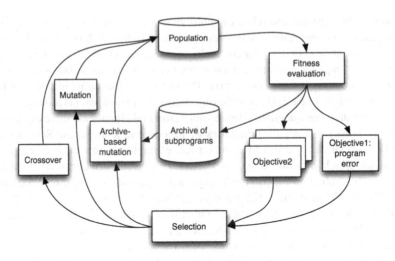

Fig. 2 As a side-effect of behavioral evaluation, evaluation can identify useful subprograms in programs being evaluated. Such subprograms can be gathered in an archive, maintained throughout the entire evolutionary run, and reused by search operators (here: archive-based mutation). Empirical evidence shows that such *code reuse* can substantially improve search performance (Krawiec and O'Reilly 2014)

configurations driven by conventional fitness functions and control configurations devised to test more specific hypotheses (e.g., which of the abovementioned search drivers is more essential for performance). In the case of tree-based GP, we also extended the approach with *code reuse*: the subprograms indicated as potentially valuable in the process (i.e., corresponding to the attributes used by a decision tree) were retrieved from the evaluated programs, stored in a carefully maintained archive, and reused by an appropriately designed mutation operator. Code reuse lead to further dramatic boosts of performance, measured in terms of success rate, error rate, predictive accuracy, and, interestingly, program size. For instance, on the suite of 35 benchmarks used in Krawiec and O'Reilly (2014), the average rank on success rate was 2.43 for PANGEA with code reuse, compared to 3.10 for conventional GP working with ten times larger population, and 3.86 for GP working with same-sized population (100). Two other PANGEA-based setups, one of them using only two objectives and the other one without archive, ranked third and fourth with average ranks of 3.36 and 3.43, respectively. Two-objective GP working with program error and program size as objectives came last, with the average rank of 4.83. Other performance indicators, like program error and predictive accuracy, were also in favor of behavioral approach. For detailed account on experimental results, see Krawiec and Swan (2013) and Krawiec and O'Reilly (2014).

4 Consequences of Behavioral Perspective

Complete characterization of program behavior can be a natural means for assessing and controlling the *diversity* of programs. For instance, a selection operator can be easily designed that, given two programs that pass the same number of tests but vary in execution record, allows them co-exist in an evolving population (by, e.g., selecting them both). No dedicated mechanism for controlling or inducing diversity may be necessary—behavioral evaluation *implicitly* provides for phenotypic diversity. This property may help mitigate the risk of premature convergence and overfocusing on local optima. The positive experimental evidence on the performance of behavioral approaches (Sect. 3.2) can be in part attributed to this characteristic. The importance of behavioral diversity has been also corroborated by methods like implicit fitness sharing (Smith et al. 1993; McKay 2000), co-solvability (Krawiec and Lichocki 2010), or more recently lexicase selection (Helmuth et al. 2015), where the last one seems to be particularly effective at trading-off diversity maintenance and selective pressure on an evolving population (Liskowski et al. 2015).

Behavioral characterization of programs may also facilitate *task decomposition*. Automatic detection of a task's internal modularity and performing appropriate decomposition has been for long considered one of the main challenges in designing intelligent systems, and is an important area of research in computational and artificial intelligence (Watson 2006). In behavioral program synthesis, there are at least two alternative avenues to decomposition, both of which can be conveniently explained by means of the execution record.

Firstly, by providing a separate account of program execution *for every test*, execution records open the door to 'horizontal', 'test-wise' task decomposition. This capability is essential also for semantic GP (and indeed other traditional approaches), where some crossover operators combine the behaviors of parents *on particular tests*. This is most evident for exact geometric semantic crossover, especially for the Boolean domain. That operator, when applied to parent programs p_1, p_2, generates a random Boolean subprogram p_r and produces an offspring that combines p_1, p_2 with p_r in a straightforward expression $(p_1 \wedge p_r) \vee (p_2 \wedge \overline{p_r})$. In the offspring, p_r works as a mask: it 'mixes' parents' behaviors by deciding, for each test individually, which parent to copy the output from. For the tests for which p_r returns *true*, the offspring behaves like p_1, and if p_r returns *false*, it behaves like p_2.

The presence of complete execution traces in an execution record also facilitates the less obvious 'vertical' task decomposition. What we mean here is the stage-wise structure of a task, as explained on the example of calculating the median in Sect. 3. In that example, the desired decomposition consists of splitting the original programming task into two separate subtasks of (1) sorting the list and (2) retrieving the central element of the sorted list. Arguably, solving each of these subtasks separately can be expected to be easier than synthesizing a complete program that calculates the median. We posit that such desired decompositions can be, at least for some programming tasks, automatically derived from a working population of programs by analyzing execution records. In Krawiec (2012), we provided some experimental evidence for this hypothesis: 'behavioral trajectories' tend to cluster, thereby revealing the internal structure of a task.

In this chapter, we considered methods that use behavioral information primarily to *drive* the selection process: an alternative evaluation function characterizes (possibly in a multi-objective fashion) the candidate solutions, and that information is used to select the most promising of them. The above remarks on task decomposition point to the alternative ways of exploiting behavioral information, in particular by redefining search operators. The code reuse mutation operator described in Sect. 3.2 and in Krawiec and O'Reilly (2014) is an example of such functionality. However, that operator implants the valuable code fragments in the offspring at random locations. Given execution records of mutated/recombined programs, search operators can be even more sophisticated in behavioral terms. For instance, a behaviorally-aware crossover operator could recombine the parents so as to achieve the desired behavioral effect (e.g. by combining a list-sorting subprogram with a subprogram that retrieves the central element from a list in the median problem mentioned earlier).

The behavioral perspective adopted in this chapter in the context of GP has interesting implications beyond program synthesis. One can draw immediate parallels between the trace of stepwise execution of a GP program on a fitness case and the search trajectory of a metaheuristic solver acting on a problem instance. The 'state' of a metaheuristic could of course also include other variables of relevance. For example, the state of Simulated Annealing would include current temperature. In the PANGEA approach described above, a search driver is induced (via a decision tree) from the executable structure. The essential difference in the extension to metaheuristics is that with the GP approach, the executable structure *is* the candidate solution, whereas in this extended approach it is the *means* by which solutions are found (i.e. the particular way in which temperature is modified throughout a Simulated Annealing run). It may nonetheless be possible to obtain search drivers in this more general context by correlating solver state against the candidate solutions representing its current search progress, using any of the gamut of ML techniques mentioned above.

There is much emphasis in the optimization research community on providing solutions for individual problem instances which are 'good enough, quickly enough'. It must not be forgotten that the most significant improvements have arisen from analytical and scientific activity, rather than the engineering activity of 'manual tweaking' of operators and parameters. It is therefore vital to build tools to help distinguish 'universal' features of solvers from 'parochial' ones. The primary strength of GP above other regression techniques is as a model-agnostic mechanism for knowledge discovery. A wider research agenda towards 'robot scientists' (Sparkes et al. 2010) that actively seek correlates between effectors (e.g. generated metaheuristic search operators) and their observed effects allows these strengths to be directed back into the optimization process itself. This wider agenda of an autonomous search agent capable of metacognitive activity invites contribution from areas such as developmental robotics (Lungarella et al. 2003) and pattern theory (Grenander 1989). This is of course a different class of activity from optimizing an individual problem instance, but architectures of this general nature (e.g. Swan et al. 2014) are necessary in order to automate that which currently requires the labour of skilled researchers.

5 Conclusions

The behavioral perspective on program synthesis urges us to rethink the structure and workflow of typical GP algorithm and generic evolutionary methods. A typical iterative optimization algorithm can be visualized as a loop of evaluation phase, selection phase, and the phase of applying search operators ('variation' in evolutionary terms). An evaluation function is typically externalized as a separate component, and communicates only with selected stages of the loop. For the behavioral approach 'in the large', it may be more appropriate to visualize the workflow as a *network of interconnected components* that exchange information about the search process. By having access to behavioral characteristics of candidate solutions, the components in such an architecture would be more empowered when making decisions about the fate of particular candidate solutions.

To an extent, the behavioral approach can be seen as a means for making search process more 'intelligent' while keeping it relatively ignorant about the domain-specific aspects. By observing program behavior as captured in an execution record, a search algorithm gains better insight into program specifics, while abstracting from characteristics of the underlying program representation, programming language, etc. For instance, PANGEA may observe similar or even the same execution records whether the evolving programs implement imperative or functional programming paradigms.

The fascinating realization is that there are probably many potentially useful search drivers beyond the conventional ones, and beyond the ones discussed in this chapter. It is even possible to that some of them may provide better performance of search algorithms than anything known to date. In this chapter and previous works on this topic, we have only scratched the surface of how search drivers can be defined [or automatically derived from a problem (Kocsis and Swan 2014)]. In a longer-term research perspective, it would be highly desirable to come up with a principled approach to the design of search drivers.

Acknowledgements Krzysztof Krawiec acknowledges support from grant 09/91/DSPB/ 0572 and National Science Centre grant 2014/15/B/ST6/05205. Una-May O'Reilly acknowledges support from Li Ka Shing Foundation.

References

Brameier M, Banzhaf W (2007) Linear genetic programming. Genetic and evolutionary computation, vol XVI. Springer, New York. http://www.springer.com/west/home/default?SGWID=4-40356-22-173660820-0

Deb K, Pratap A, Agarwal S, Meyarivan T (2002) A fast and elitist multiobjective genetic algorithm: NSGA-II. IEEE Trans Evol Comput 6(2):182–197. doi:10.1109/4235.996017

Grenander U (1989) Advances in pattern theory. Ann Stat 17(1):1–30. doi:10.1214/aos/1176347002. http://dx.doi.org/10.1214/aos/1176347002

Helmuth T, Spector L, Matheson J (2015) Solving uncompromising problems with lexicase selection. IEEE Trans Evol Comput 19(5):630–643. doi:10.1109/TEVC.2014.2362729. http://ieeexplore.ieee.org/stamp/stamp.jsp?tp=&arnumber=6920034

Hofstadter DR (1979) Godel, Escher, Bach: an eternal golden braid. Basic Books, New York

Jin Y, Olhofer M, Sendhoff B (2002) A framework for evolutionary optimization with approximate fitness functions. IEEE Trans Evol Comput 6:481–494

Kocsis ZA, Swan J (2014) Asymptotic genetic improvement programming with type functors and catamorphisms (extended abstract). In: Johnson C, Krawiec K, Alberto Moraglio MO (eds) Semantic methods in genetic programming (SMGP) at parallel problem solving from nature (PPSN XIV), Ljubljana

Krawiec K (2012) On relationships between semantic diversity, complexity and modularity of programming tasks. In: Soule T, Auger A, Moore J, Pelta D, Solnon C, Preuss M, Dorin A, Ong YS, Blum C, Silva DL, Neumann F, Yu T, Ekart A, Browne W, Kovacs T, Wong ML, Pizzuti C, Rowe J, Friedrich T, Squillero G, Bredeche N, Smith SL, Motsinger-Reif A, Lozano J, Pelikan M, Meyer-Nienberg S, Igel C, Hornby G, Doursat R, Gustafson S, Olague G, Yoo S, Clark J, Ochoa G, Pappa G, Lobo F, Tauritz D, Branke J, Deb K (eds) GECCO '12: Proceedings of the fourteenth international conference on Genetic and evolutionary computation conference. ACM, Philadelphia, pp 783–790. doi:10.1145/2330163.2330272

Krawiec K, Lichocki P (2010) Using co-solvability to model and exploit synergetic effects in evolution. In: Schaefer R, Cotta C, Kolodziej J, Rudolph G (eds) PPSN 2010 11th international conference on parallel problem solving from nature. Lecture notes in computer science, vol 6239. Springer, Krakow, pp 492–501. doi:10.1007/978-3-642-15871-1_50

Krawiec K, O'Reilly UM (2014) Behavioral programming: a broader and more detailed take on semantic GP. In: Igel C, Arnold DV, Gagne C, Popovici E, Auger A, Bacardit J, Brockhoff D, Cagnoni S, Deb K, Doerr B, Foster J, Glasmachers T, Hart E, Heywood MI, Iba H, Jacob C, Jansen T, Jin Y, Kessentini M, Knowles JD, Langdon WB, Larranaga P, Luke S, Luque G, McCall JAW, Montes de Oca MA, Motsinger-Reif A, Ong YS, Palmer M, Parsopoulos KE, Raidl G, Risi S, Ruhe G, Schaul T, Schmickl T, Sendhoff B, Stanley KO, Stuetzle T, Thierens D, Togelius J, Witt C, Zarges C (eds) GECCO '14: Proceedings of the 2014 conference on Genetic and evolutionary computation. ACM, Vancouver, pp 935–942. doi:10.1145/2576768.2598288. http://doi.acm.org/10.1145/2576768.2598288, best paper

Krawiec K, Solar-Lezama A (2014) Improving genetic programming with behavioral consistency measure. In: Bartz-Beielstein T, Branke J, Filipic B, Smith J (eds) 13th international conference on parallel problem solving from nature. Lecture notes in computer science, vol 8672. Springer, Ljubljana, pp 434–443. doi:10.1007/978-3-319-10762-2_43

Krawiec K, Swan J (2013) Pattern-guided genetic programming. In: Blum C, Alba E, Auger A, Bacardit J, Bongard J, Branke J, Bredeche N, Brockhoff D, Chicano F, Dorin A, Doursat R, Ekart A, Friedrich T, Giacobini M, Harman M, Iba H, Igel C, Jansen T, Kovacs T, Kowaliw T, Lopez-Ibanez M, Lozano JA, Luque G, McCall J, Moraglio A, Motsinger-Reif A, Neumann F, Ochoa G, Olague G, Ong YS, Palmer ME, Pappa GL, Parsopoulos KE, Schmickl T, Smith SL, Solnon C, Stuetzle T, Talbi EG, Tauritz D, Vanneschi L (eds) GECCO '13: Proceeding of the fifteenth annual conference on genetic and evolutionary computation conference. ACM, Amsterdam, pp 949–956. doi:10.1145/2463372.2463496

Liskowski P, Krawiec K, Helmuth T, Spector L (2015) Comparison of semantic-aware selection methods in genetic programming. In: Proceedings of the seventeenth annual conference on genetic and evolutionary computation companion, GECCO Comp (accepted)

Lungarella M, Metta G, Pfeifer R, Sandini G (2003) Developmental robotics: a survey. Conn Sci 15:151–190

McKay RIB (2000) Fitness sharing in genetic programming. In: Whitley D, Goldberg D, Cantu-Paz E, Spector L, Parmee I, Beyer HG (eds) Proceedings of the genetic and evolutionary computation conference (GECCO-2000). Morgan Kaufmann, Las Vegas, pp 435–442. http://www.cs.bham.ac.uk/~wbl/biblio/gecco2000/GP256.pdf

Miller GA (1983) Informavores. Wiley, New York, pp 111–113

Moraglio A, Krawiec K, Johnson CG (2012) Geometric semantic genetic programming. In: Coello CA, Cutello V, Deb K, Forrest S, Nicosia G, Pavone M (eds) Parallel problem solving from nature, PPSN XII (part 1). Lecture notes in computer science, vol 7491. Springer, Taormina, pp 21–31. doi:10.1007/978-3-642-32937-1_3

Popovici E, Bucci A, Wiegand RP, de Jong ED (2011) Coevolutionary principles. In: Handbook of natural computing. Springer, Heidelberg

Quinlan J (1992) C4.5: programs for machine learning. Morgan Kaufmann, San Mateo

Smith R, Forrest S, Perelson A (1993) Searching for diverse, cooperative populations with genetic algorithms. Evol. Comput. 1(2):127–149

Sparkes A, Aubrey W, Byrne E, Clare A, Khan M, Liakata M, Markham M, Rowland J, Soldatova L, Whelan K, Young M, King R (2010) Towards robot scientists for autonomous scientific discovery. Autom Exp 2(1):1. doi:10.1186/1759-4499-2-1. http://www.aejournal.net/content/2/1/1

Spector L, Robinson A (2002) Genetic programming and autoconstructive evolution with the push programming language. Genet Program Evolvable Mach 3(1):7–40. doi:10.1023/A:1014538503543. http://hampshire.edu/lspector/pubs/push-gpem-final.pdf

Swan J, Woodward J, Özcan E, Kendall G, Burke E (2014) Searching the hyper-heuristic design space. Cogn Comput 6(1):66–73. doi:10.1007/s12559-013-9201-8

Tomassini M, Vanneschi L, Collard P, Clergue M (2005) A study of fitness distance correlation as a difficulty measure in genetic programming. Evol Comput 13(2):213–239. doi:10.1162/1063656054088549

Watson RA (2006) Compositional evolution: the impact of sex, symbiosis and modularity on the gradualist framework of evolution. Vienna series in theoretical biology, vol NA. MIT Press, Cambridge. http://eprints.ecs.soton.ac.uk/10415/

Using Graph Databases to Explore the Dynamics of Genetic Programming Runs

Nicholas Freitag McPhee, David Donatucci, and Thomas Helmuth

Abstract For both practical reasons and those of habit, most evolutionary computation research is presented in highly summary form. These summaries, however, often obscure or completely mask the profusion of specific selections, crossovers, and mutations that are ultimately responsible for the aggregate behaviors we're interested in. In this chapter we take a different approach and use the Neo4j graph database system to record and analyze the entire genealogical history of a set of genetic programming runs. We then explore a few of these runs in detail, discovering important properties of lexicase selection; these may in turn help us better understand the dynamics of lexicase selection, and the ways in which it differs from tournament selection. More broadly, we illustrate the value of recording and analyzing this level of detail, both as a means of understanding the dynamics of particular runs, and as a way of generating questions and ideas for subsequent, broader study.

Keywords Graph database • Neo4j • Ancestry • Genealogy • Lexicase selection • Tournament selection

1 Introduction

It is common practice in empirical evolutionary computation (EC) research to perform a substantial number of runs, and then report a handful of aggregate statistics that summarize and (hopefully) represent the complex dynamics of those many runs. Tables present values such as mean or median best fitnesses at the end of runs, collapsing the complexities of dozens or hundreds of runs into a single number, possibly with a standard deviation or a confidence interval to give a sense of the distribution. Plots can often be more informative, showing how these numbers change over time during the runs, possibly giving a sense of the system dynamics

N.F. McPhee (✉) • D. Donatucci
Division of Science and Mathematics, University of Minnesota, Morris, MN, USA
e-mail: mcphee@morris.umn.edu

T. Helmuth
Computer Science, University of Massachusetts, Amherst, MA, USA

© Springer International Publishing Switzerland 2016   185
R. Riolo et al. (eds.), *Genetic Programming Theory and Practice XIII*,
Genetic and Evolutionary Computation, DOI 10.1007/978-3-319-34223-8_11

and the range of behaviors. These plots, however, are typically still aggregate representations that obscure or completely hide important moments that, if explored, might reveal valuable insight into the evolutionary dynamics being reported.

An alternative would be to collect, store, and analyze at least some of the rich panoply of evolutionary and genealogical events that make up the low-level details of these runs. Databases provide a natural tool for storing and accessing large data sets, but traditional relational databases are poorly suited for many of the queries that are important for genealogical analysis. In this chapter, we illustrate the use of graph databases as an alternative storage and analysis tool for evolutionary computation runs. We have previously demonstrated that graph databases can be an effective tool for analyzing complex genetic programming (GP) dynamics (Donatucci et al. 2014), which led directly to a proposed change to standard sub-tree crossover in tree-based GP (McPhee et al. 2015). Here we will use the open source Neo4j graph database tool[1] to explore data from a collection of PushGP runs (Helmuth et al. 2015a) on several problems drawn from a benchmark collection of introductory programming problems (Helmuth and Spector 2015).

Note that this is *not* going to be a presentation of "traditional hypothesis-driven research". It will be based on an *assumption*, namely that something interesting happens in these runs, and that we can learn useful things by exploring them in more detail, but the presentation will be fairly discursive, reflecting our back-and-forth experience of wrestling with the data. Our initial queries start from fairly obvious questions (e.g., "Why did we succeed here?"), but from there we engage in a dialog with data, letting the answers to early questions shape and guide our subsequent exploration. We are not presenting a tidy, sterile summary of our adventures, but the messier (but we think more informative in this context) journal of what Pickering might call our "mangle of practice" (Smith et al. 2008; Pickering 1993).

Here we explore the impact of lexicase (Spector 2012) and tournament selection on the dynamics of runs whose aim is to solve a basic software synthesis problem. In the process we are able to discover surprising and likely important properties of lexicase that suggest areas of additional exploration and indicate reasons for the substantially better performance seen when using lexicase on a variety of software synthesis problems (Helmuth and Spector 2015).

We're not the first people to recognize the potential value of exploring lineages and ancestry graphs. The HeuristicLab team has been working for several years on a set of tools to analyze at least small genealogical run histories (Burlacu et al. 2013, 2015); hopefully these exciting features will be in an upcoming release. Burlacu et al. (2013) also has an excellent survey of a variety of work that uses genealogical information in EC work; none of this, however, appears to save and analyze full genealogical histories, but instead tends to use local ancestry information for purposes such as diversity promotion. Recent work (Kuber et al. 2014) applies network theory to ancestry graphs, looking for things like cliques as a way of better

---

[1]http://neo4j.com/

understanding EC dynamics; that work is similar in spirit to this chapter, but differs in the kinds of graphs that are built and the tools used to analyze them.

Because we're going to focus on the use of graph databases, there will on occasion be avenues of exploration that we won't pursue because they would properly involve different tools. This exploration, for example, raises important questions about the relationships between parent and child genomes. These could be addressed using, e.g., difference-merge tools from software engineering, or sequence alignment tools from genomics; see, e.g., Burlacu et al. (2013) for an excellent example of this kind of analysis. We will, however, consider that beyond the scope of this chapter. A key value of our graph database results will be in providing focus for our use of those other tools, identifying key moments and individuals in the course of a run that deserve additional attention. There are thousands of potential genome comparisons to make in a single run, for example, but our graph databases analysis helps identify some of the critical individuals, crossovers, and mutations in the run, allowing us to concentrate on the steps that are likely to have mattered most.

We'll provide expanded motivation for this work in Sect. 2, and background on relevant tools and concepts in Sect. 3. In Sect. 4 we explore in some detail a successful lexicase selection run, identifying several properties of lexicase selection that distinguish it from other, more traditional selection methods. We then explore a successful tournament selection run in Sect. 5, comparing those results to the earlier lexicase results. In Sect. 6 we step back a little and look at the results of expanding some of our queries across hundreds of runs, and then wrap up with some conclusions in Sect. 7.

2 Motivation

Consider the job of a paleontologist, who regularly reconstructs not just individuals but also species and entire phylogenetic trees on the basis of a handful of teeth and bones, or even just impressions left in prehistoric mud. They rarely have DNA, so any evolutionary relationship is inherently speculative, subject to constant debate and revision. Even with detailed DNA sequences, the construction of phylogenetic trees for existing species is non-trivial.

In evolutionary computation, however, we have access to *everything*, at least in principle. We could gather every selection, every mutation, and every crossover as they play out in our systems. Yet we typically throw almost all that data away, reporting just aggregate statistics and summary plots, completely failing to take advantage of our privileged position, a position most paleontologists would presumably eye with considerable envy. Not only does this seem an inherent waste, these aggregations typically obscure critical moments in the dynamics of runs which might speak volumes if explored.

While this sort of aggregate reporting is often valuable, allowing for important comparative analysis, it typically fails to provide any sense of the *why*. Yes, Treatment A led to better aggregate performance than Treatment B—but what happened in the runs that led to that result? Any end result is ultimately the intricate combination of thousands or millions of selections, recombinations, and mutations, and if Treatment A is in some sense "better" than Treatment B, it must ultimately be because it affected all those genealogical and genetic events in some significant way, biasing them in a way that improved performance.

Unfortunately, published research rarely includes information that might shed light on these *why* events. We rarely see evolved programs, for example, or any kind of post-run analysis of those programs, and there is almost never any data or discussion of the genealogical history that might help us understand how a successful program actually came to be. Sometimes these events and details aren't included for reasons of space and time; evolved programs, for example, are often extremely large and complex, and a meaningful presentation and discussion of such a program could easily take up more space than authors have available. We suspect, however, that another reason this sort of *why* analysis often isn't reported is because it isn't done, in no small part because it's hard. As EC researchers we're in the "privileged" position of being able to collect anything and everything that happens in a run, but that's a potentially huge amount of data, and leaves us with two substantial problems: How to *store* the data, and how to *analyze* the data after it's stored. Decreasing data storage costs have done much to mitigate the first problem, but one still needs good tools to process and explore what could quickly run into terabytes of data.

Assuming one has access to the necessary storage, databases are the obvious tool for the collection of the data. Most common database tools, however, don't lend themselves to the kinds of analysis that we need in evolutionary computation work. Most relational and document-based databases, for example, require complex and expensive recursive joins to trace significant hereditary lines. In exploring the dynamics of an EC run, it may be necessary to make connections across dozens or even hundreds of generations, which simply isn't plausible with a relational database (Robinson et al. 2013). While we use Neo4j as our graph database in this work, there are numerous other graph databases that could potentially be effective tools (Wikipedia 2015a). We make no claims to have exhaustively explored the range of possible database tools for this sort of work.

3 A Little Background on Tools and Problems

This section provides some background on some of the key subjects of this work: The Neo4j graph database and its query language Cypher; the PushGP system; lexicase selection; and the replace-space-with-newline test problem.

3.1 Neo4j and Cypher

Graph databases (Robinson et al. 2013) are a relatively new database tool, where data is stored as a collection of nodes and relationships in a graph, with a specialized query language that makes it easy to ask questions about complex relationships. In our work, nodes typically represent individuals, and `:PARENT_OF` relationships capture the central genealogical connections. We store important data such as the total error as properties of individual nodes, and genetic operators as properties on `:PARENT_OF` edges.

The Neo4j query language, Cypher, allows patterns in this data to be readily extracted. A detailed description of Cypher is beyond the scope of this chapter, but Cypher's central feature is the ability to describe sub-graph patterns. The Neo4j engine can then search for subgraphs matching these patterns. Cypher also provides the ability to filter results based on properties in a manner quite similar to more traditional SQL queries.

3.2 PushGP

PushGP (Spector and Robinson 2002; Spector et al. 2005) is a stack-based genetic programming system. The details of PushGP aren't crucial for this analysis, but it is useful to know a few things:

- PushGP uses a linear genome, which is then converted into a program.
- PushGP supports a variety of *typed* stacks, with corresponding typed instructions. The `integer-add` instruction takes the top two items from the `integer` stack, adds them, and pushes the result back onto the `integer` stack.
- There is an `exec` stack which can hold blocks of instructions. This is what allows PushGP programs to loop or recurse, as pushing a block of instructions onto the `exec` causes those instructions to be executed next.

While traditionally PushGP has evolved Push programs themselves, the most recent version of PushGP instead evolves linear *Plush genomes* consisting of instructions paired with *close counts*. The Plush genomes are manipulated by genetic operators, but are translated into Push programs prior to execution. During translation, any instruction that uses code from the `exec` stack implicitly opens a code block; the close counts are natural numbers indicating how many open code blocks should be closed after a given instruction.

In the runs explored here, there are three genetic operations: Alternation, uniform-mutation, and uniform-close-mutation. Alternation is based on the earlier ULTRA operator (Spector and Helmuth 2013), and is similar to an N-point crossover in genetic algorithms. The two parent genomes are traversed from left to right, copying instructions from the source parent to the child. There's a small probability at each instruction of an alternation event, which switches which parent is being used as the instruction source. For every alternation event there's a small chance of slightly shifting the instruction location in the source parent; how much deviation is possible is controlled by an *alignment deviation* parameter. Uniform-mutation simply replaces each instruction with a randomly chosen instruction with some small probability. Uniform-close-mutation modifies each close count value with some small probability. The runs discussed here allowed for *pipelining* of genetic operators, so we might have combinations like alternation followed by uniform-mutation. For additional details and the particular parameters used in these runs see Helmuth and Spector (2015).

3.3 Lexicase Selection

Lexicase selection is a recently developed selection method for evolutionary computation in which individuals are selected by filtering the population according to performance on individual fitness cases, considered in random order (Spector 2012). Lexicase selection, when used as the parent selection method in genetic programming, has been shown to provide significant improvements in terms of problem-solving power (Helmuth et al. 2015b; Helmuth and Spector 2015).

For each parent selection event, lexicase selection (Algorithm 2) randomly orders the test cases and then removes any individuals that do not have the best performance on the first case. If more than one individual remains, then those that do not have the best performance *among those that remain* on the second case are also removed. This continues until only one individual remains and is selected, or until all cases have been used, in which case a random member of the set of remaining individuals is selected. Key properties of lexicase selection are (a) it avoids combining all errors into a single value, (b) because of the random ordering of test cases, every test case will be most important (first to be considered) at least occasionally, and (c) similarly, each pair of test cases, and each triple, etc., will be most important now and then.

Algorithm 2 Pseudocode for lexicase selection, in the context of error minimization. Here the function perf(i, p) computes the performance of program p on test case i

---

```
candidates := the entire population
cases := list of all the test cases in a random order
while |candidates| > 1 and |cases| > 0 do
    current, cases := first(cases), rest(cases)
    best_performance := min{perf(i, current) | i ∈ candidates}
    candidates := {i | i ∈ candidates ∧ perf(i, current) = best_performance}
end while
return random individual from candidates
```

3.4 Replace-Space-with-Newline

The replace-space-with-newline problem is an introductory programming bench-mark problem taken from Helmuth and Spector (2015). Here the program is given an input string and required to both (a) print the string with all the spaces replaced by newlines and (b) return an integer that is the number of non-space characters in the input string. There are 100 different training instances for this problem, each of which generates two error values: (a) the Levenshtein distance between the printed output and the target print string, and (b) the absolute difference between whatever value is on the top of the `integer` stack and the expected return value. A penalty value of 1000 is assigned for test cases that were expecting a return value but found the `integer` stack empty. For tournament selection runs, all 200 of these error values were added together to form the total error, which was used as the fitness for the individuals. For lexicase selection the errors were kept separate in an error vector of 200 values; this, as we shall see, frequently allowed individuals to be selected who did well on some test cases, but very poorly on others.

3.5 Our Data

In this chapter we explore a subset of the data collected for Helmuth et al. (2015a). In particular we have the full genealogical records for 100 runs of replace-space-with-newline using lexicase selection, and 100 runs using tournament selection with tournament size 7. In those runs, 57 of the 100 lexicase runs succeeded, i.e., an individual was discovered that had zero error on all 200 of the training cases. Tournament selection only had 13 successes out of 100 runs, so lexicase selection provides a significant advantage on this problem. Similar results in Helmuth and Spector (2015) indicate that lexicase is in fact generally much more successful than tournament selection across a broad range of software synthesis problems.

4 Lexicase, Meet Replace-Space-with-Newline

It's one thing to know that lexicase succeeds 57 out of 100 times on the replace-space-with-newline problem, but that leaves us with the crucial question of *why*? In order to study this question, we chose one successful run to explore in more detail. We're making no claims that this is a "representative" run (whatever that would even mean); it's an *interesting* run, though, and our hope is that by understanding its dynamics better we can learn useful things about both the problem and the tools we're applying. Looking at this run in some detail certainly unearthed several surprising results, and in Sect. 6 we'll expand our view by looking at some cumulative results across all 100 lexicase runs.

4.1 Working Backwards

A natural place to start our analysis is at the end of the run, when the GP system created one or more individuals that solved the problem. So we used Neo4j to find all the ancestors of any "winning" individual, i.e., an individual with a total error of zero represented than others. As we've already mentioned, individual 86:261 has 45 successful offspring, and both individuals 82:447 and 83:047 have five offspring in the graph, i.e., five offspring that were ancestors of a winning individual in generation 87. Each of these is marked in Fig. 1 with a shaded diamond.

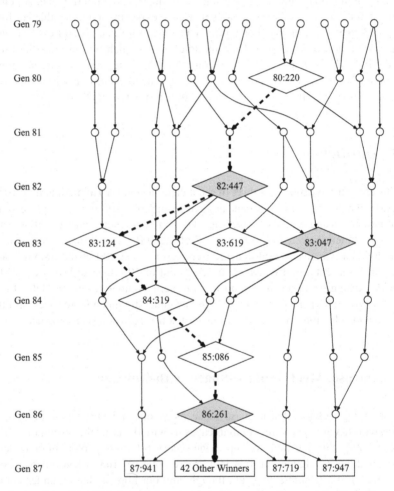

Fig. 1 Ancestry of the 45 "winners" from a successful run of replace-space-with-newline using lexicase. *Diamond-shaped* nodes had an unusually large number of offspring (over 100 each). *Shaded nodes* had at least five offspring that were ancestors of winners

Figure 1, however, only tells us how many offspring an individual had that were themselves either a winner or an ancestor of a winner, as no other nodes are displayed. One might wonder how many total offspring an individual has regardless of whether they led to a winner. Using a Cypher query to identify the most fecund ancestors of winners in these last nine generations reveals several things that were quite surprising. The most remarkable of these was that individual 86:261 was a parent of 934 of the 1000 individuals in generation 87! Given that lexicase selection was designed in significant part to spread selection events out across the population, this makes it clear that there are times when lexicase does the opposite, and instead puts nearly all its eggs in a single basket. This level of selection focus would simply be impossible using almost any other common type of selection such as tournament selection; in most uses of tournament selection, for example, no individual can be in more than a relative handful of tournaments, and thus can't be a parent terribly often no matter how fit they are.

While no other node in Fig. 1 has nearly as many children as 86:261 did, there are several that also had very high reproduction rates, putting them well above what would be possible with something like tournament selection. Individual 82:447, for example, had 443 offspring, including the 5 illustrated in Fig. 1. In fact there were eight individuals in Fig. 1 that have more than 100 offspring; each of these is indicated with a diamond shape. This highlights a particularly interesting ancestry chain from 80:220 through 81:691, 82:447, 83:124, 84:319, 85:086 to 86:261, marked with dashed edges in Fig. 1. With the exception of 81:691, which "only" had 17 offspring, each of these seven individuals had more than 100 offspring, and thus had a fairly dominate role in shaping that part of the evolutionary process

If we look at the total error in of the individuals in Fig. 1, we again find some surprises that tell us quite a lot about lexicase selection. In particular, if we look at the total error for each individual along the dashed path from 80:220 through 82:447 to 86:261, the total errors of the first five individuals in the chain are reasonably low. One (individual 82:447) has the best total error in that generation and all but 81:691 (the individual with only 17 offspring) are in the top fifth of the population when ranked by total fitness. The fitnesses of the last two (the grandparent and parent of *every* one of the 45 solutions), however, came as quite a shock. In particular, individual 85:086 has a total error of 100,000, placing it *very near the bottom of the population by total error* (rank 971). Individual 86:261, which was the parent of 924 of the 1000 individuals in the next generation, has a total error of 4034, placing it below 3/4 of the population in its generation by that aggregate measure.

How could individuals with such terrible total fitness end up being selected so often as parents? Exploring the specific test case errors reveals that individual 85:086 is perfect on half of the test cases (all those that involve printing), but gets a penalty error of 1000 on the other half because it never actually returns a value. Every one of its ancestors in Table 1, however, has at least a few non-zero errors on the printing test cases, meaning that any lexicase ordering that places a few key printing test cases before any of the "return" test cases would likely select individual 85:086.

Table 1 The total error and rank (by total error) in the population in that individual's generation for the sequence of "diamond" individuals from in Fig. 1

Individual	Total error	Rank in population
80:220	321	147
81:691	441	268
82:447	107	1
83:124	157	85
84:319	240	188
85:086	100,000	971
86:261	4034	765

What about individual 86:261, with it's 934 offspring? It has error zero on 194 of the 200 test cases. On 4 of the remaining 6 test cases it, like individual 85:086, fails to return a value and gets the penalty of 1000; it has an error of 17 on the other two. Thus it gets 97 % of the test cases correct, but happens to be *heavily* penalized for its behavior on 4 of the 6 it gets wrong. In a system that aggregates the errors, its rank of 765 out of 1000 would mean that it would probably have no offspring. With lexicase selection, however, it's success on the 194 test cases means that it is selected (from this population) almost every time. In fact only 152 of the 1000 individuals in the final generation had a parent who *wasn't* 86:261, and only 116 other individuals in generation 86 had an offspring in the next generation. While four of those had 10 or more offspring in the last generation, none of those four actually gave rise to a winner. The three parents of winners other than 86:261 (individuals 86:272, 86:049, and 86:672 in Fig. 1) had very few offspring (1, 2, and 2 respectively), suggesting that they may not have contributed much (or anything) to their successful progeny, and the success of their offspring was due more to the good fortune of mating with 86:261 than anything else.

4.2 How Exactly Did We Get Here?

Now that we know quite a lot about who gave rise to those 45 winners, what genetic operations brought them about? The largest group was 18 of the 45 which came about through uniform-close-mutation alone, *all* of which were mutations of individual 86:261. This indicates that success could be achieved via a fairly simple modification to 86:261's genome that only modifies where some code blocks end.

The other large group was 17 winners that arose via alternation followed by uniform-mutation. 14 of these were the result of a self-cross of 86:261 and itself, with the other three being crosses between 86:261 and the other three parents of winners (86:272, 86:049, and 86:672). There were also two smaller groups of winners, 6 which were the result of alternation alone (all self-crosses of 86:261), and 4 from uniform-mutation alone applied to 86:261.

An obvious question then is what changed in moving from 86:261 to the final solutions. The genomes and programs involved are fairly complex (over 200 instructions) and, as mentioned earlier, a full analysis of the genomes and behaviors of the individuals involved is beyond the scope of this chapter. Such an analysis is possible, however, and our graph database work has clearly identified individuals whose genomes and programs deserve additional study.

Based on this exploration, we can also propose a hypothesis for further exploration. 86:261's total error of 4034 comes in large part from failing to return a value on four test cases. A distinct possibility is that 86:261 simply times out on those four test cases. The efficacy of uniform-close-mutation suggests that there might be some sequence of instructions that are being executed repeatedly via a loop or recursion, and there are uniform-close-mutations that shorten that block in ways that allow it to complete all the test cases within the time limit without changing the value returned.

5   How Is Tournament Selection Different?

In addition to studying lexicase selection, we wanted to collect data from replace-space-with-new-line with tournament selection in order to compare with our earlier lexicase results. As noted in Sect. 4, lexicase produced at least one individual with an error of zero in 57 of 100 runs while tournament selection only produced 13 of 100 successful runs. In this section we'll explore one of these 13 successful tournament runs in a little more detail.

An immediate difference between the lexicase and tournament runs is that there was only one solution discovered in the tournament selection run, in contrast to the 45 different individuals that solved the problem in the lexicase run.

Figure 2 shows the ancestry of the winning individual from generation 150 (when the winner was discovered) back to generation 145. It's clear that the branching factor in this ancestry is much higher than with lexicase in Fig. 1. Table 2(a) shows the number of ancestors n generations back that contributed to the winning individual, and we can see that the number of ancestors increases much more quickly for tournament than lexicase; at 10 generations back, there were approximately three times the number of contributing parents in tournament as in lexicase. This is likely partially due to the fact that in lexicase some parents produced a surprisingly large number of children. Another possible contribution to this asymmetry is a difference in the role of mutations under lexicase, but we haven't yet explored that in any detail.

Another major difference was the selection pressure exerted by the two selection mechanisms. As we saw earlier, in lexicase selection one parent can dominate the selection if it performs well for a significant number of test cases. However, tournament selection can never impose such a strong selection pressure. Throughout the entire run, the most a single parent in the tournament selection run ever produced was 24 children (see Table 2(b)), and all of the 18 most prolific parents produced between 17 and 24 offspring. Compare this to lexicase selection, where all of the 18 top parents produced over 200 offspring. This extreme difference in selection pressure may also help explain the differences in the branching factor of the two ancestry trees.

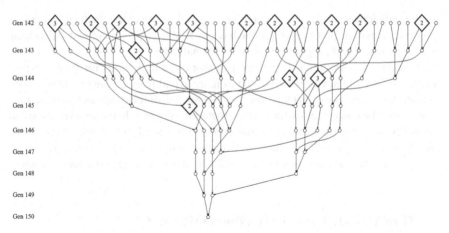

Fig. 2 Ancestry of the sole "winner" from run 74 of tournament selection, replace-space-with-newline. The few nodes with more than one offspring that is an ancestor of the winner are marked with *diamonds* containing the number of children (in this graph) for that node. Most of those nodes had additional children, not pictured in the figure, that are not ancestors of the winning individual

We also noticed another crucial difference between the types of individuals selected for reproduction. With tournament selection, the primary bias is towards individuals that have the lowest total error. However, this is not the case in lexicase where, as long as an individual performs extremely well for enough cases, it is still possible to be selected for reproduction, even if it has substantial errors on other test cases. In this tournament run, for example, every ancestor of the winner in the last six generations has a total error of either 83 or 132, which is in marked contrast to the diversity of total errors in the lexicase run (see Table 1). Additionally, across all individuals chosen as parents in the last 20 generations of the tournament run (regardless of whether they were an ancestor of the winner), there were as few as one and at most five distinct total errors within each generation. This suggests that tournament selection kept mutating and recombining a small set of behaviors until it managed, essentially by accident, to produce an improved child. Lexicase, on the other hand, maintained a much more diverse population and appeared to somehow leverage that diversity to continue to discover improvements.

6 A Few Cumulative Results

The bulk of this chapter has focused on exploring two specific successful runs on the replace-space-with-newline problem, one using lexicase selection, and one using tournament selection. To better understand how well this application of graph databases scales, we also created two larger cumulative databases (one for lexicase selection and one for tournament selection), each containing the complete genealogical record for all 100 runs on replace-space-with-newline. Given these cumulative

Table 2 Two examples of the impact of selection on evolutionary dynamics in the two explored runs

	(a)				(b)	
	Number of ancestors				Number of children	
n	Lexicase	Tournament	Rank in run	Lexicase	Tournament
18	58	297	1	934	24
17	52	236	2	657	23
16	46	180	3	594	23
15	49	152	4	590	21
14	45	209	5	433	20
13	46	212	6	326	20
12	41	146	7	297	19
11	29	97	8	294	19
10	22	63	9	285	19
9	14	42	10	283	18
8	14	33	11	279	18
7	10	30	12	271	18
6	9	20	13	234	18
5	7	13	14	220	18
4	6	10	15	212	18
3	7	6	16	205	18
2	6	4	17	203	18
1	4	2	18	202	17

Table (a) lists the number of parents contributing to a winning individual n generations away for both the lexicase and tournament runs explored in this chapter. The top row, for example, indicates that in the lexicase run there were 58 distinct ancestors of a winning individual 18 generations before the discovery of a winner, and in the tournament run there were 297 distinct ancestors 18 generations before the discovery of a winner. Table (b) lists the 18 most fecund individuals across the entirety of each of the lexicase and tournament selection runs

databases, we were then able to do broad queries against those collections of runs. These were typically inspired by observations from the explorations of individual runs, with the broader queries helping us understand to what degree an observation in an individual run was representative or an outlier.

An obvious question, for example, is how unusual is the individual we discovered in Sect. 4 that had 934 offspring? Was that an aberration, or are these kinds of hyper-selected and hyper-fecund individuals a regular occurrence when using lexicase selection? Querying the combined database revealed that there were 71 individuals in the 100 lexicase runs that were selected more than 900 times, where the average number of selections in a given generation was 1700. So each of these 71 individuals received over half the total selections in its generation, and consequently had numerous offspring; all had over 700 offspring out of the 1000 created for the next generation. 22 of those 71 individuals had over 900 offspring, with the biggest winners being two individuals that had 990 and 991 offspring, respectively, after being selected over 1600 times each.

These 71 individuals clearly represent a very small fraction of the over 18 million nodes encapsulated in our 100 lexicase runs. 50 of the 100 runs, however, had at least one individual with over 900 selections, so this kind of hyper-selection is clearly common in the dynamics of these lexicase runs. This sort of hyper-selection has a profound impact on the dynamics of a run, as almost every individual in the subsequent generation is a child of the hyper-selected individual, and due to self-crosses and mutations that individual is often the *only* parent of those children. Thus the genetics of that individual are likely to have an enormous influence on the make-up of the next generation, creating a substantial population bottleneck. So while those 71 individuals only represent a tiny proportion of the cumulative population, they're likely to have a tremendous impact on the run dynamics; thus the ability to identify and examine these individuals is potentially very informative.

One of the other surprises from our earlier exploration is how "unfit" some of those highly selected individuals were when viewed through the lens of total error. Turning now to these cumulative results, we find that 15 of these 71 hyper-selected individuals had total error at or below 10, and so would likely be selected by tournament selection (although never more than a few dozen times). On the other end of the spectrum, however, 7 of these 71 hyper-selected individuals had total error over 3000 and would have been *extremely* unlikely to ever be chosen using tournament selection. So here again we see a substantial difference between the dynamics of lexicase and tournament selection, especially given the impact these hyper-selected individuals have on their runs.

Finally, looking at all 200 runs makes it clear that lexicase and tournament selection differ considerably in the likelihood of discovering multiple "winning" individuals in the same generation. Over the 100 runs of the replace-space-with-newline with tournament selection, only 13 runs found a solution with zero total error, and only one of those runs had more than one solution in the final generation (there were two). Of the 57 successful lexicase runs, however, 30 (so just over half) had multiple solutions. Many were only a few (6 runs just had 2 solutions), but 6 runs had over 30 solutions, including runs with 69 and 74 solutions. This strongly suggests that when tournament discovered a winning individual, that discovery was fairly random and therefore had a low probability. The prevalence of multiple solutions in the lexicase runs, however, indicates that the discovery of those solutions had a much higher probability. What's less clear is whether that increased probability was driven by lexicase's hyper-selection in the last generation, or whether lexicase selection throughout the run had led to Push program structures that were easier to combine/mutate into winning individuals.

7   So What Did We Learn In All This?

In this chapter we've illustrated the potential of graph databases such as Neo4j as tools for exploring and analyzing some of the rich matrix of low-level events that ultimately make up any evolutionary computation run. Here we've used graph databases as a new kind of "lab bench" instrument, allowing us to zoom in and capture the details of potentially crucial moments in our GP runs.

This has exposed surprising and potentially important properties of lexicase selection. These properties will hopefully help us better understand dynamics of runs using lexicase, and why those runs are frequently more successful than runs using tournament selection. A key goal in the original design of lexicase selection (Helmuth et al. 2015b) was to increase and maintain diversity in GP runs which, as Helmuth et al. (2015a) shows, does appear to be the case, with lexicase generally leading to more diversity than either tournament selection or implicit fitness sharing. One would, however, typically consider hyper-selection and the consequent population bottlenecks to be the enemy of diversity, so our discovery of a pattern of individuals being selected hundreds, and sometimes even *thousands*, of times was at the very least striking and worthy of further study.

A key risk in this sort of narrow examination is that one might mistake an observation for a pattern. Thus there will always be a need for broader summary statistical analysis. The kind of exploration we've demonstrated here will ultimately have to be used in conjunction with those statistical tools; discoveries in "digs" such as these can raise questions and suggest hypotheses that can then be supported or refuted through the use of more "traditional" studies.

One other concern is how to scale our use of graph databases. Databases for individual runs are quite tractable, and can be populated and explored using basic off-the-shelf computers. Combing multiple runs, however, presents a number of challenges. The database containing the 100 lexicase runs, for example, contained over 18 million nodes and over 25 million edges, with the full Neo4j database weighing in at 18 GB. The database combining the 100 tournament runs was even larger because most of the runs went the full 300 generations: over 28 million nodes, over 48 million edges, and a database of over 31 GB. That said, these databases runs reasonably on stock desktop hardware, which is how all the results presented here were generated. To scale up further, however, we'd quickly start to need more specialized infrastructure to, for example, combine the lexicase and tournament selection runs into a single database containing the history of those 200 runs, or combining the results of runs on multiple problems. We have data from thousands of runs (Helmuth et al. 2015a), but bringing that together in a single database is clearly infeasible using these tools. An alternative to this approach would be to have a host of database engines on a cluster, each serving a subset of related data, and then providing tools that would allow queries to be run across all these endpoints, aggregating those results into a single response. Version 1.1 of the SPARQL query language (Wikipedia 2015b), for example, includes support for queries across multiple endpoints and could potentially be used to run queries across large distributed datasets.

Acknowledgements Thanks to the members of the Hampshire College Computational Intelligence Lab and M. Kirbie Dramdahl at the University of Minnesota, Morris, for discussions that helped to improve the work described in this chapter. Thanks also to Josiah Erikson for systems support, and to Hampshire College for support for the Hampshire College Institute for Computational Intelligence. This material is based upon work supported by the National Science Foundation under Grants No. 1017817, 1129139, and 1331283. Any opinions, findings, and conclusions or recommendations expressed in this publication are those of the authors and do not necessarily reflect the views of the National Science Foundation.

We are very grateful to all the participants in the 2015 Genetic Programming Theory and Practice (GPTP) workshop for their enthusiasm, ideas, and support. In particular we'd like to thank William Tozier for all manner of suggestions and feedback, and in particular for helping us understand the connection between our work and the Pickering's idea of the "mangle of practice". Krzysztof Krawiec provided a number of valuable suggestions based on an early draft. Steven Gustafson suggested that we look into SPARQL and triplestore databases as an alternative to Neo4j, an interesting idea we haven't had time to explore in detail. Stuart Card connected us to the interesting related work by Karthik Kuber. Finally, thanks to the GPTP organizers; without their hard work none of those other valuable conversations would have occurred.

References

Burlacu B, Affenzeller M, Kommenda M, Winkler S, Kronberger G (2013) Visualization of genetic lineages and inheritance information in genetic programming. In: GECCO '13 Companion: proceeding of the fifteenth annual conference companion on genetic and evolutionary computation conference companion. ACM, Amsterdam, pp 1351–1358

Burlacu B, Affenzeller M, Winkler S, Kommenda M, Kronberger G (2015) Methods for genealogy and building block analysis in genetic programming. In: Computational intelligence and efficiency in engineering systems, studies in computational intelligence, vol 595. Springer International Publishing, Berlin, pp 61–74

Donatucci D, Dramdahl MK, McPhee NF (2014) Analysis of genetic programming ancestry using a graph database. In: Proceedings of the Midwest Instruction and Computing Symposium. http://goo.gl/RZXY2U

Helmuth T, Spector L (2015) General program synthesis benchmark suite. In: Silva S, Esparcia-Alcazar AI, Lopez-Ibanez M, Mostaghim S, Timmis J, Zarges C, Correia L, Soule T, Giacobini M, Urbanowicz R, Akimoto Y, Glasmachers T, Fernandez de Vega F, Hoover A, Larranaga P, Soto M, Cotta C, Pereira FB, Handl J, Koutnik J, Gaspar-Cunha A, Trautmann H, Mouret JB, Risi S, Costa E, Schuetze O, Krawiec K, Moraglio A, Miller JF, Widera P, Cagnoni S, Merelo J, Hart E, Trujillo L, Kessentini M, Ochoa G, Chicano F, Doerr C (eds) GECCO '15: Proceedings of the 2015 on genetic and evolutionary computation conference. ACM, Madrid, pp 1039–1046. doi:10.1145/2739480.2754769, http://doi.acm.org/10.1145/2739480.2754769

Helmuth T, McPhee NF, Spector L (2015a) Lexicase selection for program synthesis: a diversity analysis. In: Riolo R, Worzel WP, Groscurth K (eds) Genetic programming theory and practice XIII. Springer, Ann Arbor; Genetic and Evolutionary Computation, forthcoming

Helmuth T, Spector L, Matheson J (2015b) Solving uncompromising problems with lexicase selection. IEEE Trans Evol Comput 19(5):630–643. doi:10.1109/TEVC.2014.2362729, http://ieeexplore.ieee.org/stamp/stamp.jsp?tp=&arnumber=6920034

Kuber K, Card SW, Mehrotra KG, Mohan CK (2014) Ancestral networks in evolutionary algorithms. In: Proceedings of the 2014 conference companion on Genetic and evolutionary computation companion. ACM, pp 115–116

McPhee NF, Dramdahl MK, Donatucci D (2015) Impact of crossover bias in genetic programming. In: GECCO '15: Proceedings of the 2015 conference on genetic and evolutionary computation

Pickering A (1993) The mangle of practice: agency and emergence in the sociology of science. Am J Sociol 99(3):559–589. http://www.jstor.org/stable/2781283

Robinson I, Webber J, Eifrem E (2013) Graph Databases. O'Reilly, URL http://info.neotechnology.com/rs/neotechnology/images/GraphDatabases.pdf

Smith BH, Weintraub ER, Franklin A, Pickering A, Guzik K (2008) The mangle in practice: science, society, and becoming. Duke University Press, Durham/London

Spector L (2012) Assessment of problem modality by differential performance of lexicase selection in genetic programming: a preliminary report. In: McClymont K, Keedwell E (eds) 1st workshop on understanding problems (GECCO-UP), ACM, Philadelphia, pp 401–408. doi:10.1145/2330784.2330846, http://hampshire.edu/lspector/pubs/wk09p4-spector.pdf

Spector L, Helmuth T (2013) Uniform linear transformation with repair and alternation in genetic programming. In: Riolo R, Moore JH, Kotanchek M (eds) Genetic programming theory and practice XI. Genetic and Evolutionary Computation, chap 8. Springer, Ann Arbor, pp 137–153

Spector L, Robinson A (2002) Genetic programming and autoconstructive evolution with the push programming language. Genet Program Evolvable Mach 3(1):7–40. doi:10.1023/A:1014538503543, http://hampshire.edu/lspector/pubs/push-gpem-final.pdf

Spector L, Klein J, Keijzer M (2005) The push3 execution stack and the evolution of control. In: GECCO 2005: Proceedings of the 2005 conference on Genetic and evolutionary computation, vol 2. ACM, Washington D.C., pp 1689–1696

Wikipedia (2015a) Graph database — Wikipedia, the free encyclopedia. http://en.wikipedia.org/w/index.php?title=Graph_database&oldid=653752823. Accessed 28 March 2015, Online

Wikipedia (2015b) SPARQL — Wikipedia, the free encyclopedia. http://en.wikipedia.org/w/index.php?title=SPARQL&oldid=661970711. Accessed 8 June 2015, Online

Predicting Product Choice with Symbolic Regression and Classification

Philip Truscott and Michael F. Korns

Abstract Market researchers often conduct surveys to measure how much value consumers place on the various features of a product. The resulting data should enable managers to combine these utility values in different ways to predict the market share of a product with a new configuration of features. Researchers assess the accuracy of these choice models by measuring the extent to which the summed utilities can predict actual market shares when respondents choose from sets of complete products. The current paper includes data from 201 consumers who gave ratings to 18 cell phone features and then ranked eight complete cell phones. A simple summing of the utility values predicted the correct product on the ranking task for 22.8 % of respondents. Another accuracy measurement is to compare the market shares for each product using the ranking task and the estimated market shares based on summed utilities. This produced a mean absolute difference between ranked and estimated market shares of 7.8 %. The current paper applied two broad strategies to improve these prediction methods. Various evolutionary search methods were used to classify the data for each respondent to predict one of eight discrete choices. The fitness measure of the classification approach seeks to reduce the Classification Error Percent (CEP) which minimizes the percent of incorrect classifications. This produced a significantly better fit with the hit rate rising from 22.8 to 35.8 %. The mean absolute deviation between actual and estimated market shares declined from 7.8 to 6.1 % (p. <0.01). A simple language specification will be illustrated to define symbolic regression and classification searches.

Keywords Abstract regression grammars • Genetic algorithms • Symbolic regression • Classification • Non-linear regression

P. Truscott (✉)
Southwest Baptist University, Bolivar, MO 65613, USA
e-mail: ptruscott@sbuniv.edu

M.F. Korns
Analytic Research Foundation, 2240 Village Walk Drive Suite 2305, Henderson,
NV 89052, USA
e-mail: mkorns@korns.com

© Springer International Publishing Switzerland 2016
R. Riolo et al. (eds.), *Genetic Programming Theory and Practice XIII*,
Genetic and Evolutionary Computation, DOI 10.1007/978-3-319-34223-8_12

1 Introduction

Market research has a long history of attempting to evaluate the importance of product features so that brand managers can predict the popularity of new feature combinations. Some of these methods require respondents to consider a set of products and place them into a rank ordering. Since these survey methods require the respondents to assess all of a product's features jointly the methodology has been termed 'Conjoint'(Green and Rao 1971). After the respondents have ranked the complete configurations the utility values of the individual product features are calculated using multinomial logit.

Another approach requires respondents to explain how much value they place on each feature separately. This "self-explicated" approach generates utility values directly (Marder 1997).

Both conjoint and self-explication methods then incorporate the utility values into predictive models. The utilities are re-combined to play "what-if" games that predict the market share of future product offerings. The most common technique for assessing the accuracy of these choice models is to follow the conjoint or self-explication survey task with a validation task. Often these validation tasks resemble the process of using comparison-shopping Web sites. In "full profile" validation tasks the entire feature matrix of the products are displayed so the interplay of different feature combinations will be apparent in a way that is unlikely during non-Internet purchasing. The validation task constructed for the current research only displayed the brand and model number of each product and thus avoided "leading the witness" by giving prominence to specific features.

The goal is to find a methodology that is able to predict a high proportion of the top products in the validation task. A high "hit rate" substantiates the accuracy of a given methodology.

Codd (1983) made a major contribution to the standardization of database searches through the specification that came to be known as the Structured Query Language (SQL). It is argued here that a standardization of evolutionary searches would be similarly beneficial. For this reason, the following article illustrates a parsimonious specification of four of the most common type of classification searches: classification and regression trees (CART), neural networks, decision trees and non-linear discriminant analysis.

2 The Experiment

For the current research, a self-explication survey collected data from 201 Indian consumers. The specific form of rating has been termed the Un-bounded Write-in Scale (UWS) because respondents may give rating numbers without upper or lower limits (Marder 1997). A Web page tells them to click:

Table 1 Illustrative mobile phone attributes

Diagonal Screen Size	Price	Operating System
Less than 3 Inch	5000 Rupees or less	Android
3.0 - 3.4 Inches	5001 - 10000 Rupees	Symbian
3.5 - 3.9 Inches	10001 - 18000 Rupees	Windows
4.0 - 4.4 Inches	18001 - 35000 Rupees	Blackberry
4.5 - 4.9 Inches	35001 Rupees and Above	iOS (iPhone OS)
5 Inches and over		

1. a plus sign button as many times as they want to show how much they like a feature
2. a minus sign button as many times as they want to show how much they dislike a feature
3. a zero button to indicate that they are neutral about a feature

The chief proponent of this methodology, Marder (1997), argues that the resultant ratings are superior to bounded ratings because they lead to normal distributions. Ratings scales, for example, that limit choices from one to ten often produce "cliff distributions" where the values cluster at the minimum or maximum value. For the current research respondents were required to evaluate 18 attributes of a mobile phone. Table 1 shows three of the 18 attributes. As illustrated by Table 1 the "attributes" of a product include both its physical features (for example screen size and CPU speed) and non-physical qualities (such as its warrantee duration, price category, and brand). The attributes did not all have the same number of levels. As can be seen from Table 1 the screen size attribute had six levels, while price had only five. To complete the Web based survey respondents were required to enter ratings for every level of every attribute. The full list of attributes and levels is shown in Appendix 1. After giving their ratings to the separate product features, respondents saw a product-ranking screen. The lower left part of the screen contained a list of eight mobile phones. The survey software randomized the list in a different way for each respondent. The lower right hand side of the screen showed an empty list which respondents were required to fill. They had to rank the eight products from "least likely to buy" to "most likely to buy". The software prevented respondents from proceeding to the final screens until they had completed the ranking task. After completing the survey, respondents were sent electronic money in the form of a 500 Indian Rupee electronic gift certificate.

3   Results from Utility Summation

The characteristics of the eight mobile phones were researched to find their actual feature configurations. Appendix 2 shows the sources for the product feature information. Table 2 shows implied market shares based on the two types of data

Table 2 Market shares from direct choice task and utility summation

iPhone 5 with 32 GB	Samsung Galaxy Note 2	Black-berry Curve 9220	XOLO Q1000	Spice Mi-495	Micro-max Canvas 4 A210	Nokia Lumia 520	Lava Iris 504Q	Prediction Method
22.4%	50.2%	2.0%	6.0%	3.0%	6.5%	9.5%	0.5%	Summed Utilities
41.3%	21.4%	6.0%	4.0%	2.5%	6.5%	12.9%	5.5%	Ranking Task
18.9%	-28.9%	4.0%	-2.0%	-0.5%	0.0%	3.5%	5.0%	Difference
Mean Absolution Deviation: 7.8%								
Hit Rate: 22.9%								
N: 201								

collected. The first row shows the number of products that achieved the highest score based on summing the utilities for the 18 attributes of the various products. Formula 1 below describes the scoring process that determines the market shares in row 1 of Table 2.

$$PV_{pi} = \sum_{a=1}^{amax} U_{ai} * F_{ap} \tag{1}$$

In formula 1 above, U_{ai} is the utility value of the ith respondent for the ath attribute level. F_{ap} represents a matrix describing the configuration of a specific product. It contains Boolean $(0,1)$ values that indicate which level of a given attribute a product has. To take the example of our mobile phone survey, if the largest screen size takes the value '1' and the iPhone has this screen size then it will take the value '1' for this level and all other levels of the screen size attribute will be zero. F_{ap} is the presence of feature F for the attribute-level a for product p. The total product value, PV_{pi}, is the sum of a product's utilities for the ith respondent for product p (given the feature configuration F_p).

Row 2 of Table 2 shows the proportion of respondents who put each of the products at the top of the list in the 'probability to purchase' ranking task. The mean absolute deviation between the estimated and direct choice markets shares was 7.8 %. The summed utility method predicted the top ranked products for 22.8 % of the respondents (a proportion commonly called the 'hit rate' in market research literature).

4 Fitness Measures and Classification Problems

For non-logit regression models, predicted values are continuous variables. The evolutionary search process results in sets of such variables that minimize the error between their predicted values and those of the dependent variable. For classification searches, the predicted values are categorical variables. The predicted values for our product search process are discrete categorical values between one and eight that represent one of the eight products.

The classification search required a suitable database of training data. The 18 utility scores were the independent variables. Each of the 201 respondents had eight sets of utility scores to represent the eight products in the ranking task. For each person the eight rows of product data were based on:

1. the utility score the respondent gave to each of the 18 attributes
2. the specific utility value that was relevant to each product's feature configuration

The evolutionary search was conducted using Abstract Regression Classification (ARC) software (Korns 2011, 2007, 2010). The dependent variable was the rank order number of the eight products where the number zero represented the product the respondents were least likely to buy and seven was the product they were most likely to buy.

The data in Table 2 GPTP appears to be an ideal candidate for a classification: GPTP search. The independent variables are 18 product feature utilities for each of the eight products. The dependent variables are eight discrete values that represent one of the eight products. This requires a fitness measure to replace the Normalized Least Squares Error (NLSE) commonly used in regression models where the dependent variable is a quantitative variable. For this reason, the current classification search used ARC's Classification Error Percent (CEP) fitness measure. This minimizes the percent of observations where the prediction was not an exact match.

5 The Select() Command

Since the classification search requires a prediction in the form of discrete values, several of the goal specifications below needed a command to constrain predicted values to be in this form. For this reason a *select()* command was used to transform continuous results into one of eight values representing one of the eight products. The following example illustrates the use of the select command. The neuralnet command (described below) can be specified to produce a certain number of outputs. In the illustrative command below the final numeric parameter (the number 8) specifies that the neuralnet goal will have eight outputs.

$$neuralnet(0,18,4,8,n)$$

In order to constrain this goal to produce a discrete value between one and eight it was embedded within a select command as follows:

$$select(neuralnet(0,18,4,8,n))$$

The select() command will analyze the vector of eight output values and return the position of the highest value. The resulting value was an integer from one to eight.

6 Training and Testing Data

Initial training runs used data for all respondents, however this meant using the records for those products that received a lower (2nd and below) choice rankings. This process resulted in fitness scores close to a level that would be produced by chance. Since classification models could expect to produce random hits 12.5 % of the time, CEP error levels close to 87.5 % were similar to the results of chance.

An alternative search strategy involved using only the data records for the top ranked products during the training stage, but then applying the resulting model to the full data set during testing. This procedure was followed in the searches described below.

7 A Decision Tree Search

A form of decision tree searching is described in Breiman et al. (1984) where the predicted outcome variable is a category. This form of their search process has been termed a classification tree (as distinction from a regression tree described below). A tree search can be specified in ARC using the 'tree' code-expression generator in the following form:

tree(categories, node-depth, tree-depth, c |v |f)

The final parameter takes the following values:

1. 'c' signifies that there is a constant at the decision node
2. 'v' signifies that there is an abstract variable at the decision node
3. 'f' signifies that there is a function at the decision node

In the case of our cell phone search task the goal was specified as follows:

model(tree(8,2,3,f))

Thus, eight categories were specified. The node-depth was two. The tree depth was three and functions were at the decision node. After running for 3 h and evaluating 142,000 formulas, the champion formula produced the data in Table 3. In terms of the metrics used by the market research industry, the product hit rate worsened from 22.8 % under summed utilities to only 2.2 %. The Mean Absolute Deviation between actual and estimated choice shares also deteriorated from 7.8 to 21.4 %.

Table 3 Market shares from direct choice task and a decision tree search

iPhone 5 with 32 GB	Samsung Galaxy Note 2	Black-berry Curve 9220	XOLO Q1000	Spice Mi-495	Micro-max Canvas 4 A210	Nokia Lumia 520	Lava Iris 504Q	Prediction Method
0.0%	0.0%	0.0%	0.0%	30.5%	23.8%	0.0%	45.7%	Decision Tree
41.3%	21.4%	6.0%	4.0%	2.5%	6.5%	12.9%	5.5%	Ranking Task
-41.3%	-21.4%	-6.0%	-4.0%	28.0%	17.4%	-12.9%	40.2%	Difference
Mean Absolute Deviation: 21.4%								
Hit Rate: 2.2%								
N:201								

8 A Non-Linear Discriminant Analysis (NLDA) Search

The next evolutionary search involved Linear Discriminant Analysis (LDA). Since this search was conducted at a node-depth of two (see below) this was technically Non-Linear Discriminant Analysis (NLDA). The 'net' code-expression generator for LDS searches takes the following form:

$$net(node\text{-}depth, inputs, outputs, x \mid v, n \mid h \mid s)$$

The penultimate parameter was introduced to handle extremely large numbers of input variables. Its two values have the following meanings:

1. 'x' signifies concrete features (when there are fewer than 250 independent variables)
2. 'v' signifies abstract variables (when there are more than 250 independent variables)

The final parameter allows the user to constrain the output to be in one of three forms:

1. 'n' signifies 'no operator' (results unconstrained)
2. 'h' signifies hyperbolic tangent (results in the range -1 to $+1$)
3. 's' signifies sigmoid (results in the range 0 to 1)

The specific goal for the product search was:

$$select(net(2,18,8,x,n))$$

Two represented the node-depth. The 18 utility scores were the inputs, and the eight outputs corresponded to the eight product choices. The 'x' parameter implies concrete features rather than abstract variables. The final parameter indicates the output values were unconstrained but since the select command was wrapped around the goal specification, the outputs were constrained to be in the range of 1–8. After 24,000 well-formed formulas, the NLDA search produced a champion on the testing data with a CEP error of 92 % (Table 4).

Table 4 Market shares from direct choice task and non-linear discriminant analysis

iPhone 5 with 32 GB	Samsung Galaxy Note 2	Black-berry Curve 9220	XOLO Q1000	Spice Mi-495	Micro-max Canvas 4 A210	Nokia Lumia 520	Lava Iris 504Q	Prediction Method
7.8%	1.2%	32.1%	0.0%	0.0%	16.9%	0.0%	41.9%	LDA
41.3%	21.4%	6.0%	4.0%	2.5%	6.5%	12.9%	5.5%	Ranking Task
-33.5%	-20.1%	26.1%	-4.0%	-2.5%	10.4%	-12.9%	36.4%	Difference
Mean Absolute Deviation: 18.3%								
Hit Rate: 8.0%								
N: 201								

Table 5 Market shares from direct choice task and weighted search

iPhone 5 with 32 GB	Samsung Galaxy Note 2	Black-berry Curve 9220	XOLO Q1000	Spice Mi-495	Micro-max Canvas 4 A210	Nokia Lumia 520	Lava Iris 504Q	Prediction Method
11.4%	10.7%	1.9%	0.0%	0.0%	5.2%	24.4%	46.4%	Weighted
41.3%	21.4%	6.0%	4.0%	2.5%	6.5%	12.9%	5.5%	Ranking Task
-29.9%	-10.7%	-4.1%	-4.0%	-2.5%	-1.3%	11.5%	40.9%	Difference
Mean Absolute Deviation: 13.1%								
Hit Rate: 14.%								
N: 201								

9 A Weighted Search

The weighted() command differs from the net() command above in that it does not guarantee coverage of all the features. The net() command guarantees coverage due to the deterministic nature of the search. Due to this determinism, all the independent variables must be included in every evolution of the formula. The general form of the weighted code-expression generator is:

$$\text{Weighted (node-depth, base-functions, n |h |s)}$$

The final parameter has the same meanings as described above under the LDA search section. The specific form of the weighted search used for the mobile phone search was:

$$\text{model(select(weighted(5,8,s)))}$$

This implied a node-depth of five, eight base functions and outputs constrained to be in sigmoid form. After evaluating 23,000 formulas this search produced the champion associated with Table 5. Compared to the NLDA search the hit rate and the difference between actual and estimated choice shares improved. However, both were still worse than the simple process of summing utilities shown in Table 2. The champion model had a CEP error of 48 %.

Table 6 Market shares from direct choice task and neural net search

iPhone 5 with 32 GB	Samsung Galaxy Note 2	Black-berry Curve 9220	XOLO Q1000	Spice Mi-495	Micro-max Canvas 4 A210	Nokia Lumia 520	Lava Iris 504Q	Prediction Method
17.3%	13.0%	0.0%	0.0%	6.8%	3.1%	6.2%	53.5%	Neural Net
41.3%	21.4%	6.0%	4.0%	2.5%	6.5%	12.9%	5.5%	Ranking Task
-24.0%	-8.4%	-6.0%	-4.0%	4.4%	-3.4%	-6.7%	48.1%	Difference
Mean Absolute Deviation: 13.1%								
Hit Rate: 15.6%								
N: 201								

10   An Artificial Neural Network (ANN) Search

McCulloch and Pitts (1943) proposed that neural events and relationships could be represented by propositional logic. Since then various algorithms have been proposed to mimic neural activity that fall into the class of Artificial Neural Networks (ANN).

ARC's neural net code-generator can create a classification goal as follows:

neuralnet(node-depth, inputs, hidden, outputs, x lv, n lh ls)

The penultimate parameter (x lv) has the same meaning as in the case of the LDA search above. The final parameter values (n lh ls) have the same meanings as they do in LDA goal specification. The specific form of the goal for the product search was:

select(neuralnet(0,18,4,8,n))

This indicates a node-depth of zero. The 18 inputs were the 18 product feature utility variables. There were four hidden layers. Eight output values represented the eight product choices. The select command wrapped around this goal constrained these outputs to be integers from one to eight.

After evaluating 11,000 formulas this goal produced the champion with the results shown in Table 6. This champion had a CEP error of 44 %. ARC allows for a search based on the Classification and Regression Tree technique described by Breiman et al. (1984). The general form of the goal specification is:

cart(node-depth, tree-depth, c lv lf)

The final parameter takes the following values:

1. 'c' signifies that there is a constant at the decision node
2. 'v' signifies that there is an abstract variable at the decision node
3. 'f' signifies that there is a function at the decision node

The specific goal for our product classification search was:

model(cart(2,3,c))

Table 7 Market shares from direct choice task and CART search

iPhone 5 with 32 GB	Samsung Galaxy Note 2	Black-berry Curve 9220	XOLO Q1000	Spice Mi-495	Micro-max Canvas 4 A210	Nokia Lumia 520	Lava Iris 504Q	Prediction Method
91.5%	0.0%	7.5%	1.0%	0.0%	0.0%	0.0%	0.0%	CART
41.3%	21.4%	6.0%	4.0%	2.5%	6.5%	12.9%	5.5%	Ranking Task
50.2%	-21.4%	1.5%	-3.0%	-2.5%	-6.5%	-12.9%	-5.5%	Difference
Mean Absolute Deviation: 12.9%								
Hit Rate: 37.8%								
N: 201								

This the goal specified a node-depth of 2 at the leaf level, a tree-depth of 3 and constants at the decision node. Since the select command was not used, the results are not constrained to be continuous values. This goal produced a regression model with a continuous variable as its predicted values.

The CART search evaluated 988,000 formulas. Its training score is not comparable to the fitness percentages above because the output from CART is a regression formula rather than a category. Its error of 97 % appears to be larger than the fitness percent errors quoted above but it produced the best metrics in terms of hit rate and mean absolute deviation between actual and estimated choice shares.

It is interesting to note that the hit rate is an improvement on that based on summed utilities in Table 2 but the Mean Absolute Deviation is worse (Table 7).

11 An NLSE Search

The favorable hit rate from the CART search suggested the possibility of using a regression model search rather than classification. Normalized Least Squares Error (NLSE) was chosen as the fitness measure. For any given respondent the full data set was used in training (the winning products and the lower ranked products) because all rankings were considered to have information value in the NLSE search process.

ARC's universal code-expression generator has the following general format:

$$universal(node\text{-}depth, base\text{-}functions, v \mid t)$$

The first parameter specifies the grammar depth of the expression allowed. The second parameter specifies the number of base functions. The final parameter has the following meanings:

1. 'v' means only variables may compose the base functions
2. 't' means variables or constants may compose the base functions

The specific goal for the mobile phone search was defined as follows:

$$regress(universal(1,14,v))$$

Table 8 Market shares from direct choice task and NLSE search

iPhone 5 with 32 GB	Samsung Galaxy Note 2	Black-berry Curve 9220	XOLO Q1000	Spice Mi-495	Micro-max Canvas 4 A210	Nokia Lumia 520	Lava Iris 504Q	Prediction Method
62.7%	11.4%	9.0%	2.0%	1.5%	1.5%	11.4%	0.5%	Summed Utilities
41.3%	21.4%	6.0%	4.0%	2.5%	6.5%	12.9%	5.5%	Ranking Task
21.4%	-10.0%	3.0%	-2.0%	-1.0%	-5.0%	-1.5%	-5.0%	Difference
Mean Absolute Deviation: 6.1%								
Hit Rate: 35.8%								
N: 201								

This specified a grammar depth of one, 14 basis functions, and only variables within them. After various combinations of operators and evolution durations no champion model improved on the summed utility approach in Table 2.

It was not clear how best to balance the importance of the winning product and the lower ranked products. Since market shares depend only on a person's top-ranked product, it was attractive to privilege them in the search process. However, the lower ranked products represented seven eighths of all the available data. A hybrid approach was selected. The dependent variable was squared. Since the top ranked product had the highest value this meant its ranking had more importance but the data from the lower ranked products was retained. This eventually produced a champion that improved on both the hit rate and the mean absolute deviation between the actual and estimated choice shares. Table 8 shows the results of this champion. The Table 8 results show that the hit rate increased from 22.8 to 35.8 % and the mean absolute deviation fell from 7.8 to 6.1 %. The Bowker-McNemar test is a variation on the Chi-square test where the same respondents are measured twice. This test indicated that the NLSE champion formula not only produced improved results but that they were different from the summed utility results (p. <0.01).

The special circumstances of product choice modeling imply that the ideal search evolutionary search process would involve a customized fitness measure, which progressively decreases the mean absolute deviation between actual and estimated choice shares.

12 Summary

Given that the cell phone data involved eight discrete choices, it was logical to assume that a predictive model could follow a classification approach. It is interesting to note, that the same training data format could be used for a wide variety of different classification search strategies. CART, decision tree learning, neural nets, non-linear discriminant analysis and it non-deterministic variant 'weighted()'. Code-generators allowed these searches to be undertaken with minimal effort to specify each search goal.

Ultimately, the product prediction process was improved by a regression approach using least squares rather than classification error as the fitness measure. Even though classification proved not to win the hunt, the ease with which hunters can use different sets of dogs must be counted as one of ARC's strengths. This should be counted as an argument for the standardization of regression and classification search languages so that this flexibility becomes commonplace.

Appendix 1: Questionnaire Text

1. Operating system

   a. Android
   b. Symbian
   c. Windows
   d. Blackberry
   e. iOS (iPhone OS)

2. Screen size

   a. Less than 3 in.
   b. 3.0–3.4 in.
   c. 3.5–3.9 in.
   d. 4.0–4.4 in.
   e. 4.5–4.9 in.
   f. 5 in. and over

3. Camera memory

   a. Below 2 megapixels
   b. 2–4.9 megapixels
   c. 5–7.9 megapixels
   d. 8 Megapixels and above

4. Memory

   a. Below 8 GB
   b. 8–15.9 GB
   c. 16–31.9 GB
   d. 32–63.9 GB
   e. 64 GB or more

5. Talk time

   a. Less than 6 h
   b. 6–11 h
   c. 12–23 h
   d. 24–35 h
   e. 36 h or more

6. Stand by time

   a. Under 50 h
   b. 50–99 h
   c. 100–199 h
   d. 200–299 h
   e. 300 h or more

7. Price

   a. 5000 Rs or less
   b. 5001–10,000 Rs
   c. 10,001–18,000 Rs
   d. 18,001–35,000 Rs
   e. 35,001 Rs and above

8. Phone thickness

   a. Less than 6 mm
   b. 6–7 mm
   c. 8–9 mm
   d. 10–11 mm
   e. 12 mm or more

9. CPU speed

   a. 1 GHz or less
   b. 1.0–1.3 GHz
   c. 1.4–1.5 GHz
   d. 1.6–1.9 GHz
   e. 2.0 GHz or more

10. Warranty length

    a. Free repairs for 6 months
    b. Free repairs for 1 year
    c. Free repairs for 1.5 years
    d. Free repairs for 2 years
    e. Free repairs for 2.5 years

11. GPS

    a. Has GPS
    b. No GPS

12. Wi-Fi

    a. Has Wi-Fi
    b. No Wi-Fi

13. Touchscreen

    a. Has a touchscreen
    b. No touchscreen

14. SIM format

    a. Single SIM
    b. Dual SIM

15. 3G

    a. Has 3G connectivity
    b. No 3G connectivity

16. Qwerty keyboard

    a. Has a QWERTY keyboard
    b. No QWERTY keyboard

17. Brand impression

    a. Apple
    b. Samsung
    c. Blackberry
    d. XOLO
    e. Spice
    f. Micromax
    g. Nokia
    h. Lava

Appendix 2: Sources of Feature Data

All feature data for the eight mobile phones were drawn from www.Flipkart.com on September 26th, 2013 except the following items that were missing from the Flipkart comparison screens.

For the iPhone 5 with 32 GB, data was missing for the CPU speed attribute. This was taken from www.GSMArena.com on September 26th 2013.

For the Samsung Galaxy Note 2, data was missing for the talk-time and standby time attributes. This was taken from www.GSMArena.com on September 26th 2013.

For the Blackberry Curve 9220, data was missing for the GPS attribute. This was taken from www.GSMArena.com on September 26th 2013. The CPU speed attribute was missing from both these sources. It was taken from asia.cnet.com on September 26th 2013.

For the XOLO Q1000, data was missing for the GPS attribute. This was taken from www.GSMArena.com on September 26th 2013.

For the Spice MI-495, data was missing for the USB connection attribute. This was taken from www.GSMArena.com on September 26th 2013. The phone thickness attribute was missing from both these sources. It was taken from comapareindia.in.com on November 5th 2013.

For the Micromax Canvas 4 A210, data was missing for the GPS attribute. This was taken from www.GSMArena.com on September 26th 2013.

For the Lava Iris 504Q, data was missing for the GPS attribute. It was taken from comapareindia.in.com on November 5th 2013.

References

Breiman L, Friedman J, RAOlshen, Stone C (1984) Classification and regression trees. Wadsworth and Brooks, Pacific Grove

Codd EF (1983) A relational model of data for large shared data banks. Commun ACM 26(1): 64–69

Green PE, Rao V (1971) Conjoint measurement for quantifying judgmental data. J Mark Res 8(3):355–363

Korns MF (2007) Large-scale, time-constrained symbolic regression-classification. In: Riolo RL, Soule T, Worzel B (eds) Genetic programming theory and practice V. Genetic and evolutionary computation, chap 4. Springer, Ann Arbor, pp 53–68. doi:doi:10.1007/978-0-387-76308-8_4

Korns MF (2010) Abstract expression grammar symbolic regression. In: Riolo R, McConaghy T, Vladislavleva E (eds) Genetic programming theory and practice VIII. Genetic and Evolutionary Computation, vol 8, chap 7. Springer, Ann Arbor, pp 109–128. http://www.springer.com/computer/ai/book/978-1-4419-7746-5

Korns MF (2011) Accuracy in symbolic regression. In: Riolo R, Vladislavleva E, Moore JH (eds) Genetic programming theory and practice IX. Genetic and evolutionary computation, chap 8. Springer, Ann Arbor, pp 129–151. doi:doi:10.1007/978-1-4614-1770-5_8

Marder E (1997) The laws of choice: predicting customer behavior. The Free Press, New York

McCulloch W, Pitts W (1943) A logical calculus of the ideas immanent in nervous activity. Bull Math Biophys 5(4):115–133

Multiclass Classification Through Multidimensional Clustering

Sara Silva, Luis Muñoz, Leonardo Trujillo, Vijay Ingalalli, Mauro Castelli, and Leonardo Vanneschi

Abstract Classification is one of the most important machine learning tasks in science and engineering. However, it can be a difficult task, in particular when a high number of classes is involved. Genetic Programming, despite its recognized successfulness in so many different domains, is one of the machine learning methods that typically struggles, and often fails, to provide accurate solutions for multiclass classification problems. We present a novel algorithm for tree based GP that incorporates some ideas on the representation of the solution space in higher dimensions, and can be generalized to other types of GP. We test three variants of this new approach on a large set of benchmark problems from several different sources, and observe their competitiveness against the most successful state-of-the-art classifiers like Random Forests, Random Subspaces and Multilayer Perceptron.

Keywords Classification • Multiple classes • Clustering

1 Introduction

In the last two decades, Genetic Programming (GP) (Koza 1992) has established itself as a solid research field, not only because of the numerous practical successes that have been reported in many different application domains (Poli et al. 2008; Koza 2010) but also due to the strengthening of its theoretical foundations (Langdon and Poli 2002).

S. Silva (✉)
Faculty of Sciences, BioISI – Biosystems & Integrative Sciences Institute, University of Lisbon, Lisbon, Portugal
e-mail: sara@fc.ul.pt

L. Muñoz • L. Trujillo
Tree-Lab, Posgrado en Ciencias de la Ingeniería, Instituto Tecnológico de Tijuana, Blvd. Industrial y Av. ITR Tijuana S/N, Mesa Otay C.P. 22500, Tijuana, B.C., Mexico

V. Ingalalli
LIRMM, Montpellier, France

M. Castelli • L. Vanneschi
NOVA IMS, Universidade Nova de Lisboa, 1070-312 Lisbon, Portugal

© Springer International Publishing Switzerland 2016   219
R. Riolo et al. (eds.), *Genetic Programming Theory and Practice XIII*,
Genetic and Evolutionary Computation, DOI 10.1007/978-3-319-34223-8_13

Probably the most straightforward formulation for a GP search is to apply it in supervised machine learning problems, particularly symbolic regression and data classification. In general, for a supervised classification problem some pattern $\mathbf{x} \in \mathbb{R}^p$ has to be classified in one of M classes $\omega_1, \ldots, \omega_M$ using a training set $\mathcal{X}$ of N p-dimensional patterns with a known class label. Then, the goal is to build a mapping $g(\mathbf{x}) : \mathbb{R}^p \to M$, that assigns each pattern $\mathbf{x}$ to a corresponding class ω_i, where g is derived based on the evidence provided by $\mathcal{X}$. In these problems fitness is usually assigned in two general ways. One approach is to use a *wrapper* method, where GP is used as a feature extraction method that performs the transformation $k(\mathbf{x}) : \mathbb{R}^p \to \mathbb{R}^d$, and then another classifier is used to measure the quality of the transformation based on accuracy or another performance measure. The second approach is to use GP to evolve g directly, performing the feature transformation step implicitly.

However, various references report on the poor performance of GP in multiclass classification (i.e., where $M > 2$) when compared to other state-of-the-art classifiers (see for instance Castelli et al. 2013). Very recently, we have introduced a novel method in Ingalalli et al. (2014), and an improved variant in Muñoz et al. (2015), which has finally allowed GP to be considered as a competitive option for multiclass classification. The current work summarizes these two previous contributions, and introduces yet another variant of the method, reporting comparative results between the three of them, and also putting them against the most popular state-of-the-art classification methods.

The remainder of this chapter is organized as follows. Section 2 describes the state-of-the-art of multiclass classification with GP that is related to the work presented here. Sections 3–5 describe the three variants of the novel method mentioned above, called M2GP, M3GP and eM3GP, respectively. Section 6 specifies the data set, tools and parameters used to perform the experiments. Section 7 reports and discusses the results achieved by each of the three variants, comparing them between each other and with the state-of-the-art classifiers. Finally, Sect. 8 concludes and proposes the future directions for this work.

2 Related Work

Espejo et al. (2010) present a comprehensive discussion on GP-based classification methods. Here we outline several GP methods that have been proposed in order to specifically tackle multiclass classification problems.

Several works (Bojarczuk et al. 2000; Sakprasat and Sinclair 2007; Shen et al. 2003; Falco et al. 2002; Tan et al. 2002) in this area are based on a common and straightforward approach that consists in evolving a single rule in each GP run. In particular, c runs are performed for a c-class classification problem. In this way, the final classifier has a single rule for each class. All these works evolve multiple comprehensible IF-THEN classification rules.

However, the focus of this short literature review is on another common approach, which consists in evolving a discriminant function. In this case the two main approaches are (1) range selection methods and (2) binary decomposition methods. Range selection methods are applicable to GP classifiers that output numerical values. The method works by declaring $c - 1$ thresholds for c-class classification problems. To select optimal thresholds, several mechanisms have been proposed, including static thresholds selection (Tackett 1993; Zhang and Ciesielski 1999), dynamic thresholds (Zhang and Smart 2004; Li et al. 2007) and slotted thresholds (Zhang and Smart 2004).

In binary decomposition methods, one classifier is trained to recognize samples belonging to a particular class and reject all other samples. This results in c classifiers for a c-class classification problem. A well-known drawback of this approach is related to the fact that the multiple classifiers may result in conflicts, whose number usually grows up proportionally to the number of classes. Hence, this approach produces an increased classification error as the number of classes gets larger. Binary decomposition methods have been explored in Kishore et al. (2000), Silva and Tseng (2008), Lin et al. (2008). The two approaches for multiclass classification, constructing a single classification function or c binary classifiers, are compared in Teredesai and Govindaraju (2004), by considering a hand-written digit recognition problem. As reported in Espejo et al. (2010), when a single function is evolved, able to discriminate all the classes, the function directly outputs the numeric value of the predicted class, since each class is an integer digit. In both cases, the fitness function is based on classification accuracy.

In Muni et al. (2004) the authors proposed a GP-based approach to multiclass classification in which each individual is a multitree structure made of c trees, where c is the number of classes. Each of these c trees $(T_1, \cdots, T_c)$ encodes a threshold function for a particular class. The system considers that a data instance x belonging to class i is correctly classified if $T_i(x) \geq 0$ and $T_j(x) < 0$, for all $j \neq i$. The fitness function is computed as the classification accuracy. A similar system evolving a multiple-threshold discriminant function is described in Winkler et al. (2007), where a fitness function based on the sum of squared errors is employed.

One of the most recent contributions of GP for multiclass classification is found in Jabeen and Baig (2013). In this work, the authors propose a two-stage strategy for multiclass classification problems, which is an improvement of a traditional binary decomposition method.

Finally, we briefly address two previous works that present a similar goal to our own, highlighting the main differences to the present contribution.

Lin et al. (2007) proposed a layered multipopulation approach, where each layer has d populations, and each population produces a single transformation $k(\mathbf{x}) : \mathbb{R}^p \rightarrow \mathbb{R}$, and classification is performed based on a threshold. While each population is evaluated independently, all of them are combined to generate new feature vectors of dimension d, which are given as input to a new layer, and only the final layer has a single population with $d = 1$. For multiclass problems an Euclidean distance classifier was used and results show the method improves the search efficiency and reduces training time. However, the approach does not improve upon the performance of a standard GP classifier, it is not tested on problems with many classes (highest is $M = 3$), and it requires an a priori setting for the number of layers and populations used in each layer.

Another, more closely related work, is the one presented by Zhang and Rockett (2009), who propose a multidimensional feature extraction method that uses a similar solution representation to the one used in the present contribution. However, the authors set a fixed limit on the maximum number of feature dimensions, set to $d = 50$, and initialize the population with trees that use different number of features within this range. Other important difference is that the authors use a multiobjective search process considering class separation and solution size, and do not explicitly consider multiclass problems, instead relying on a hierarchical nesting of binary classifiers.

3 M2GP: Multidimensional Multiclass GP

The basic idea of M2GP (originally presented in Ingalalli et al. 2014) is to find a transformation, such that the transformed data of each class can be grouped into unique clusters. In M2GP the number of dimensions in which the clustering is performed is completely independent from the number of classes, such that high dimensional datasets may be easily classified by a low dimensional clustering, while low dimensional datasets may be better classified by a high dimensional clustering.

In order to achieve this, M2GP uses a representation for the solutions that allows them to perform the mapping $k(\mathbf{x}) : \mathbb{R}^p \to \mathbb{R}^d$. The representation is basically the same used for regular tree-based GP, except that the root node of the tree exists only to define the number of dimensions d of the new space. Each branch stemming directly from the root performs the mapping in one of the d dimensions. The genetic operators are the regular subtree crossover and mutation, except that the root is never chosen as the crossing or mutation node. However, the truly specialized element of M2GP is the fitness function. Each individual is evaluated in the following way:

1. All the p-dimensional samples of the training set are mapped into the new d-dimensional space (each branch of the tree is one of the d dimensions).
2. On this new space, for each of the M classes in the data, the covariance matrix and the cluster centroid is calculated from the samples belonging to that class.
3. The Mahalanobis distance between each sample and each of the M centroids is calculated. Each sample is assigned the class whose centroid is closer. Fitness is the accuracy of this classification (the percentage of samples correctly classified).

Figure 1 shows an example of clustering of a dataset. The original data, regardless of how many features, or attributes, it contains, is mapped into a new 3-dimensional space by a tree whose root note has three branches, each performing the mapping on each of the three axes X, Y, Z. The fact that the data contains three classes is purely coincidental—it could contain any number of classes, regardless of the dimension of the space. On the left, the clustering was obtained by an individual with low accuracy; on the right, the same data clustered by an individual with accuracy close to 100 %. The class centroids are marked with large circles.

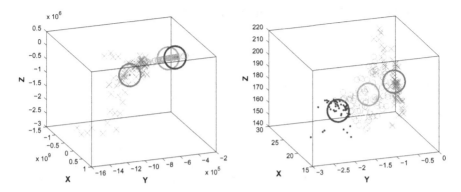

Fig. 1 Example of clustering of a dataset. On the *left*, clustering obtained by an individual with low accuracy; on the *right*, the same data clustered by an individual with very high accuracy. The *large circles* represent the centroids

At the end of the run, the solution given to the user is composed not only of the tree of the best individual, but also of the respective covariance matrices and cluster centroids. In order to classify unseen data, M2GP uses the tree to map the new samples into the new space, and then uses the covariance matrices and the cluster centroids in order to determine the minimum Mahalanobis distance between each sample and each centroid. (Note that the covariance matrices and cluster centroids are not recalculated when classifying new data.)

The choice of the Mahalanobis distance instead of the Euclidean distance is not an unnecessary complication of the algorithm. Preliminary results have consistently shown that the distance measure indeed plays a significant role in the performance of M2GP, especially in the higher dimensional solution spaces. Unlike the Euclidean distance, the Mahalanobis distance not only is able to capture the physical distance between the sample and the class clustered data sets, but also considers the statistical correlation between them, thereby reasserting the work of Shiming Xiang and Zhang (2008).

However, M2GP suffers from a drawback: how to choose the right number of dimensions for a given problem? M2GP is incapable of adding or removing dimensions during the evolution, so the number of dimensions d is fixed in the beginning of each run. M2GP chooses d based on the observation that the best fitness found on the initial generation is highly correlated with the best fitness found on the final generation (Ingalalli et al. 2014). Therefore, before initiating a run, M2GP runs a procedure that iteratively initializes different populations with increasing dimensions (we mean the dimension d mentioned earlier, not the number of individuals in the population) and checks which of these initial populations has the best fitness. Starting with $d = 1$, this procedure adds one more dimension and initializes one more population as long as the best fitness continues to improve from the previous population. As soon as adding one more dimension degrades the fitness, the procedure stops and the dimension yielding the best fitness is chosen.

4   M3GP: M2GP with Multidimensional Populations

As described in the previous section, the original M2GP uses a greedy approach to determine how many dimensions the evolved solutions should have. It may happen that by fixing the number of dimensions in the beginning of the run, the algorithm is being kept from finding better solutions during the search, ones that may use a different number of dimensions. Therefore, the newer algorithm (originally presented in Muñoz et al. 2015) evolves a population that may contain individuals of several different dimensions. It is called M3GP, which stands for M2GP with multidimensional populations. The genetic operators may add or remove dimensions, and it is assumed that selection will be sufficient to discard the worst ones and maintain the best ones in the population. The next paragraphs describe specific relevant aspects of M3GP.

4.1   Initial Population

M3GP starts the evolution with a random population where all the individuals have only one dimension. This ensures that the evolutionary search begins looking for simple, one dimensional solutions, before moving towards higher dimensional solutions, which might also be more complex.

For M2GP, a Ramped Half-and-Half initialization (Koza 1992) skewed to 25 % Grow and 75 % Full was recommended by Ingalalli et al. (2014), suggesting that a higher proportion of full trees facilitates the initial evolution. Because all the initial M3GP individuals are unidimensional, it makes sense to believe that the need for bigger initial trees is even higher. Therefore, all the individuals in the initial M3GP population are created using the Full initialization method (Koza 1992). Additionally to the Full initialization, there was also an attempt to use deeper initial trees of depth 9 instead of 6. However, preliminary results did not show any improvement, and therefore the traditional initial depth of 6 levels was used.

4.2   Mutation

During the breeding phase, whenever mutation is the chosen genetic operator, one of three actions is performed, with equal probability: (1) standard subtree mutation, where a randomly created new tree replaces a randomly chosen branch (excluding the root node) of the parent tree; (2) adding a randomly created new tree as a new branch of the root node, effectively adding one dimension to the parent tree; and (3) randomly removing a complete branch of the root node, effectively removing one dimension from the parent tree.

As mentioned previously, M3GP begins with a population that only contains unidimensional individuals. From here, the algorithm has to be able to explore several different dimensions. In M3GP mutation is the only way of adding and removing dimensions, and therefore we have increased its probability of occurrence from 0.1 (used by M2GP in Ingalalli et al. 2014) to 0.5, to guarantee a proper search for the right dimension. Preliminary results have confirmed that a higher mutation rate indeed improves the fitness.

4.3   Crossover

Whenever crossover is chosen, one of two actions is performed, with equal probability: (1) standard subtree crossover, where a random node (excluding the root node) is chosen in each of the parents, and the respective branches swapped; (2) swapping of dimensions, where a random complete branch of the root node is chosen in each parent, and swapped between each other, effectively swapping dimensions between the parents. The second event is just a particular case of the first, where the crossing nodes are guaranteed to be directly connected to the root node.

4.4   Pruning

Mutation, as described above, makes it easy for M3GP to add dimensions to the solutions. However, many times some of the dimensions actually degrade the fitness of the individual, so they would be better removed. Mutation can also remove dimensions but, as described above, it does so randomly and blind to fitness. To maintain the simplicity and complete stochasticity of the genetic operators, we have decided not to make any of them more 'intelligent', and instead we remove the detrimental dimensions by pruning the best individual after the breeding phase.

The pruning procedure removes the first dimension and reevaluates the tree. If the fitness improves, the pruned tree replaces the original and goes through pruning of the next dimension. Otherwise, the pruned tree is discarded and the original tree goes through pruning of the next dimension. The procedure stops after pruning the last dimension.

Pruning is applied only to the best individual in each generation. Applying it to all the individuals in the population could pose two problems: (1) a significantly higher computational demand, where a considerable amount of effort would be spent on individuals that would still be unfit after pruning; (2) although not confirmed, the danger of causing premature convergence due to excessive removal of genetic material, the same way that code editing has shown to cause it (Haynes 1998).

Preliminary experiments have revealed that pruning the best individual of each generation shifts the distribution of the number of dimensions to lower values (or prevents it from shifting to higher values so easily) during the evolution, without harming fitness.

4.5 Elitism

It was mentioned earlier that, in order to explore solutions of different dimensions, M3GP relies on mutation to add and remove dimensions from the individuals, with a fairly high probability. It also has to rely on selection to keep the best dimensions in the population and discard the worst ones. The way to do this is by ensuring some elitism on the survival of the individuals from one generation to the next. M3GP does not allow the best individual of any generation to be lost, and always copies it to the next generation. Let us recall that this individual is already optimized in the sense that it went through pruning. Preliminary experiments have shown that elitism is indeed able to improve fitness.

5 eM3GP: M3GP Ensemble Classifier

M3GP assumes that a single transformation will simplify the classification problem for all the classes. However, this may not be the case. It may happen that the optimal data transformation is in fact class dependent, i.e., different data clusters require transformations that change the geometrical distribution of the data points in specialized ways.

Another problem with M3GP seems to be the automatically chosen number of dimensions. In most problems, the number of dimensions used by M3GP is much larger than what M2GP uses, even when the performance on the test set is statistically equivalent (see Table 3 in Sect. 7.2). For instance, a notorious example is the WAV dataset, where the median test accuracy is almost the same for M2GP and M3GP (84.9 and 84.3, respectively) but the median number of dimensions used in the population is quite different (5 and 31, respectively). This suggests that M3GP may be suffering from bloat at the dimension level.

Finally, like many other classifiers, M3GP appears to suffer from overfitting, and is negatively affected by class imbalance, two issues that need to be addressed in real-world scenarios.

To address these issues, we propose an ensemble method called ensemble M3GP, or simply eM3GP, whereby classification is done using M different transformations, one for each class in an multiclass problem with M classes. The proposed eM3GP uses basically the same methods and representation scheme as M3GP, with the following enhancements.

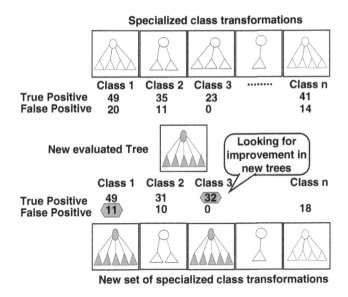

Fig. 2 Identification of the best transformation for each class based on true and false positives

1. **Specialized class transformations:** First, we are interested in identifying the best transformation k_i' for each class ω_i, and building a set of specialized class transformations $S = (k_1', \ldots, k_M')$. At the beginning of the search, when the first individual is evaluated, M copies are stored in S. For all subsequent transformations evolved during the search, we compute the number of True Positives (TP) and False Positives (FP) it produces for each class. If at least one of these numbers improves (TP gets higher or FP gets lower) and neither value deteriorates relative to k_i', then the new tree replaces k_i' within S. This is done for every individual evaluated during the run—see Fig. 2 for an illustration of this process.

2. **Ensemble classifier:** Our proposal builds an ensemble $E = (e_1, \ldots, e_M)$ of transformations for each new individual k, by combining k with the transformations stored in S. Each e_i represents a d-dimensional transformation, such that given a data point x the e_i transformation is used to compute the distance to the i-th class cluster. After computing the distance to each class cluster, the minimum distance determines the class label assigned.

3. **Ensemble construction:** One possible approach is to use S as an ensemble, however preliminary tests using this approach have shown a substantial decrease in test performance. Therefore, an ensemble E is constructed for each individual tree k and used to assign fitness. First, every element e_i in the ensemble is set equal to k, i.e., $e_i = k$ for all i. Then, the accuracy of E is computed. Afterwards, in a random order we replace each e_i by its corresponding specialized

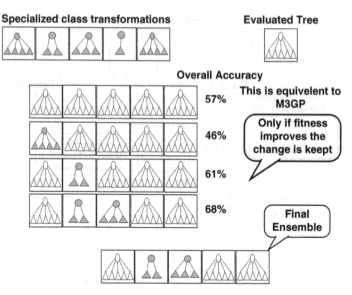

Fig. 3 Ensemble construction to build the individual

transformation k_i' in S, i.e., $e_i = k_i'$. After each replacement, we compute the accuracy of the ensemble E. If the accuracy improves then the change is kept, otherwise it is reversed. The process is depicted in Fig. 3.

There are several comments to be made regarding the proposed algorithm. First, specialized class transformations are chosen based on the performance achieved on each class, attempting to find improvements in terms of both TP and FP. These criteria provide a robust estimate of performance on a class by class basis, however it is possible that in the end we do not have the best possible transformation for each class, but only a non-dominated individual of a larger Pareto set. Nonetheless, we feel this selection process provides a useful first approximation. Second, we can say that the proposed ensembles are used to construct improved versions of each individual. This should give low quality individuals a chance to improve, and possibly save any useful genetic material they may have. Finally, the ensemble construction process is a greedy algorithm that may not be considering higher order epistatic effects. Again, for now we choose the simplest approach, and will leave future improvements as possible future research.

6   Experimental Setup

In this section we describe the data sets used for testing the methods, as well as the tools and parameters adopted for performing the experiments.

Table 1 Data sets used for the experimental analysis

Data set	HRT	IM-3	WAV	SEG	IM-10	YST	VOW	M-L
No. of classes	2	3	3	7	10	10	11	15
No. of attributes	13	6	40	19	6	8	13	90
No. of samples	270	322	5000	2310	6798	1484	990	360

6.1 Data Sets

We have used eight different data sets to test the performance of the three methods. Table 1 summarizes the main characteristics of these data sets that encompass both real-world and synthetic data, having integer and real data types, with varying number of attributes, classes and samples. The 'Heart' (HRT), 'Segment' (SEG), 'Vowel' (VOW), 'Yeast' (YST) and 'movement-libras' (M-L) data sets can be found in the KEEL dataset repository[1] in Alcala-Fdez et al. (2011), whereas the 'Waveform' (WAV) data set is available in Bache and Lichman (2013). 'IM-3' and 'IM-10' are the landsat satellite data sets that were used in Ingalalli et al. (2014) and Muñoz et al. (2015), taken from data available on the U.S. Geological Survey (USGS) Earth Resources Observation Systems (EROS) Data Center (EDC).[2] None of the eight data sets have missing values. From each of the original datasets we have formed 30 different partitions with the training and test data ratio of 70:30, to be used in 30 independent runs.

6.2 Tools and Parameters

A modified version of GPLAB 3 was used to execute all the runs. GPLAB is a freely available open source GP toolbox for MATLAB.[3] Most of the settings adopted were the GPLAB 3 defaults. The population size was 500 individuals, allowed to evolve for 100 generations in 30 independent runs per experiment. The function set included $+$, $-$, $\times$ and $/$ (protected as in Koza 1992) and the terminal set included ephemeral random constants (also as in Koza 1992). Due to the implementation particularities and differences between M2GP and M3GP, some relevant settings were modified accordingly, as already described in Sect. 4. For additional details on other settings, the reader is referred to Ingalalli et al. (2014) and Muñoz et al. (2015).

---

[1] http://keel.es/datasets.php

[2] http://glovis.usgs.gov

[3] http://gplab.sourceforge.net

For the comparison with the state-of-the-art classifiers, we have used Weka 3.6.10. Weka is also open source and freely available.[4] In Weka we have used the default parameters and configurations for each algorithm.

7 Results and Discussion

This section is split in three parts. First we summarize the results of the comparison between M2GP and a standard GP classifier, and between M2GP and a number of state-of-the-art classifiers (previously published in Ingalalli et al. 2014). Then we summarize the results of the comparison between M3GP and M2GP, and between M3GP and the best state-of-the-art classifiers from the first part (previously published in Muñoz et al. 2015). Finally, we present the new results obtained with eM3GP, comparing them with the ones obtained by the previous methods.

7.1 Results of M2GP

With the goal of comparing the performance of M2GP with the performance of other GP systems, we chose the range selection method with static threshold selection mentioned in Sect. 2 (Zhang and Smart 2004; Li et al. 2007) as the benchmark for comparison, since it is a fairly standard way of performing multiclass classification with GP. However, in data sets with a higher number of classes we immediately observed the often reported inadequacy of this standard GP method to perform multiclass classification. It was losing the race too quickly, so we abandoned any further comparison. Just to provide some numbers, on the WAV and SEG data sets M2GP improved the accuracy upon the standard method in approximately 25 and 55 percentual points, respectively.

We then compared M2GP with a number of classifiers available in Weka. Random Forests (RF) and Decision Trees (J48) are tree based classifiers; Random Subspace (RS) and Multi-Class Classifier (MCC) are meta classifiers; Multilayer Perceptron (MLP) and Support Vector Machines (SVM) are function based classifiers. Table 2 presents the results already reported in Ingalalli et al. (2014), the median and the best accuracy values of the 30 different runs for the test data sets. We have used the same set of 30 different partitions to perform 30 different runs with all the classifiers. M2GP used 100 generations and the dimension d was automatically chosen during the process of initialization (as explained in Sect. 3), except for the binary class data set (HRT) where $d = 1$, since this was reported to be the best setting. For the rest of the classifiers, we have used the default settings from Weka. SVM used the "one-against-one" approach to multi-

---

[4]http://www.cs.waikato.ac.nz/ml/weka

Table 2 Comparison among various classifiers

→ Data Set ↓ Classifiers		HRT C=2	IM-3 C=3	WAV C=3	SEG C=7	IM-10 C=10	YST C=10	VOW C=11	M-L C=15
SVM	Median	*55.556*	93.814	**86.3**	*55.844*	90.363	*41.124*	81.818	*14.352*
	Best	*65.432*	*97.938*	88.067	*61.616*	92.055	46.067	85.859	*24.074*
J48	Median	79.630	93.814	*74.800*	96.104	94.654	55.169	75.926	63.426
	Best	85.185	**98.969**	78	97.691	95.537	57.977	83.838	75.000
RF	Median	80.247	94.845	81.500	**97.258**	**96.861**	**57.528**	**89.394**	71.759
	Best	87.654	**98.969**	83.067	**98.557**	**97.744**	61.124	93.266	76.852
RS	Median	**81.481**	92.784	82.200	95.960	93.919	56.629	82.828	65.741
	Best	90.124	*97.938*	84.400	97.403	95.096	60.674	88.216	74.074
MLP	Median	80.247	95.876	83.333	96.320	90.216	**57.977**	82.492	**75.926**
	Best	87.654	*97.938*	85.200	97.403	91.319	**62.921**	87.542	**84.259**
MCC	Median	**83.951**	95.361	**86.800**	92.424	*81.829*	**57.977**	*57.576*	60.648
	Best	**90.124**	*97.938*	**88.267**	94.228	*83.865*	62.247	*65.657*	72.222
M2GP	Median	**82.099**	94.845	84.867	95.599	90.191	53.82	85.859	62.963
	Best	88.889	**98.969**	86.467	97.403	92.545	60.225	**94.613**	74.074

Median accuracy and Best accuracy on the test data set for 30 runs are reported. For each problem, the best values among the classifiers are in bold (if more than one, it means there is no statistically significant difference between their medians) and the worst values are in italics (the same). For each problem, a highlighted (respectively underlined) value means the classifier is significantly better (respectively worse) than M2GP

class classification, which has comparable performance to "one-against-all" while requiring less training time (Hsu and Lin 2002). To test for statistical significance of the results, the non-parametric Kruskal-Wallis with Bonferroni correction was used under the alternative hypothesis that the accuracy values of the different classifiers do not have equal medians.

Table 2 has many things to reveal. First of all, on the IM-3 data set all the classifiers obtained median accuracy values that are not statistically different from each other. In terms of best accuracy, on this data set M2GP was one of the classifiers achieving the best value (in bold). Also in the VOW data set M2GP achieved the best accuracy. Regarding the median accuracy values, M2GP was one of the best classifiers on HRT (in bold), and never one of the worst classifiers on any of the data sets (in italics). On data sets WAV, YST and VOW, only the best classifiers were able to outperform M2GP (highlighted values), whereas M2GP was able to outperform many other classifiers (underlined values), at least one on each data set except IM-3. Ingalalli et al. (2014) report that on the M-L data set M2GP was not able to choose the ideal d, otherwise it would probably have been able to outperform more classifiers. Regarding the comparison with the other function based classifiers (MLP and SVM), M2GP was clearly superior to SVM in almost all problems, and fairly competitive with MLP, which together with MCC was one of the best classifiers. RF was, however, the clear winner, in particular on the data sets with a higher number of classes.

7.2 Results of M3GP

The comparison between M3GP and M2GP will be presented in terms of fitness, expressed as classification accuracy, and in terms of number of nodes and number of dimensions of the solutions. Whenever a result is said to be significantly different (better or worse) from another, it means the difference is statistically significant according to the Friedman test with Bonferroni-Holm correction using the 0.05 significance level.

Table 3 shows quantitative results regarding the training and test fitness, also including the information on the number of nodes of the best individuals, as well as their number of dimensions. All these results refer to the median of the 30 runs. The best approach (between M2GP and M3GP, and the new eM3GP whose results will be discussed later) on each problem is marked in bold—more than one is marked when the difference is not statistically significant. In terms of size, or number of nodes, we consider lower to be better. In terms of dimensions, we remark that a higher number of dimensions does not necessarily translate into a larger number of nodes and/or lower interpretability of the solutions. We include additional information for the number of dimensions, which is the minimum and maximum values obtained in the 30 runs.

Table 3 shows that, in terms of training fitness, M3GP is significantly better than M2GP in all the problems (except the last, M-L, where the results are considered the same), while in terms of test fitness M3GP is better or equal to M2GP in all problems (except M-L). It is interesting to note that it is in the higher dimensional problems (except M-L) that M3GP achieves better results than M2GP (the problems are roughly ordered by dimensionality of the data). Ingalalli et al. (2014) had already identified problem M-L as yielding a different behavior from the others, and in Muñoz et al. (2015) it was once again often considered the exception to the rule. Our explanation for M3GP not being able to perform better on this problem is the extreme easiness it has in reaching maximal accuracy. Both M2GP and M3GP achieve 100 % training accuracy, but M3GP does it in only a few generations (not shown), producing very small and accurate solutions that barely generalize to unseen data. On the other hand, M2GP does not converge immediately, so in its effort to learn the characteristics of the data it also evolves some generalization ability.

Regarding the size of the solutions, in most problems where M3GP brought improvements, it also brought larger trees. However, when we split the nodes of the M3GP trees among their several dimensions, even the largest trees (e.g., in IM-10 and YST) seem to be simple and manageable (around 20 nodes per dimension), in particular when we consider that no simplification has been done except for the pruning of detrimental dimensions (see Sect. 4), and therefore the effective size of the trees may be even smaller.

Regarding the number of dimensions used in M2GP and M3GP, two things become clear. The first one is that there seems to be no single optimal number of dimensions for a given problem, since both M2GP and M3GP may choose wildly

Table 3 Comparison between M2GP, M3GP and eM3GP

	HRT	IM-3	WAV	SEG	IM-10	YST	VOW	M-L
Training fitness								
M2GP	89.4	98.2	87.4	96.8	91.4	62.6	95.9	**100**
M3GP	**94.7**	**99.6**	**90.7**	**98.1**	**93.0**	**68.5**	**100**	**100**
eM3GP	86.7	98.2	81.8	96.1	92.0	61.0	87.8	**100**
Test fitness								
M2GP	**80.2**	**93.8**	**84.9**	**95.6**	90.2	53.8	85.9	**63.0**
M3GP	79.0	**95.4**	**84.3**	**95.6**	**91.0**	**56.2**	**93.8**	57.1
eM3GP	**80.8**	**93.2**	81.2	94.7	**90.3**	**56.1**	78.6	**65.1**
# Nodes								
M2GP	37	24	126	43	117	146	49	33
M3GP	110	66	71	111	239	274	53	13
eM3GP	4	8	3	8	58	14	10	4
# Dimensions								
M2GP	3 *(1–8)*	2 *(1–4)*	5 *(2–10)*	4 *(3–8)*	7 *(4–10)*	6 *(1–13)*	9 *(4–18)*	10 *(7–12)*
M3GP	12 *(1–17)*	5 *(2–8)*	31 *(29–37)*	11 *(5–21)*	12 *(11–16)*	13 *(11–18)*	20 *(16–20)*	12 *(10–13)*
eM3GP	1 *(1–4)*	1 *(1–5)*	1 *(1–10)*	6 *(2–10)*	7 *(3–12)*	10 *(1–16)*	4 *(1–14)*	2 *(1–11)*

The values refer the medians of 30 runs. The best values are in bold (if more than one, it means there is no statistically significant difference between the medians)

Whenever a result is said to be significantly different (better or worse) from another, it means the difference is statistically significant according to the Friedman test with Bonferroni-Holm correction using the 0.05 significance level

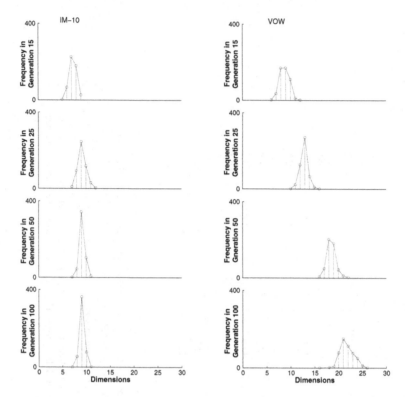

Fig. 4 Distribution of the number of dimensions in the population in generations 15, 25, 50 and 100 (*top to bottom*) for M3GP. On the *left*, a typical run of problem IM-10. On the *right*, a typical run of problem VOW

different values, depending on the run. The second one is that M3GP tends to use a larger number of dimensions than M2GP. What these numbers do not show is that different problems result in very different behaviors with respect to the evolution of the number of dimensions. Figure 4 illustrates two main types of behavior, described next. In most problems the distribution of the number of dimensions moves rapidly to higher values in the beginning of the run, and then remains stable and more or less in the same range until the end of the run (exemplified on the left in Fig. 4). However, in some problems, like WAV and VOW, the distribution of the number of dimensions does not settle during the 100 generations of the run, and instead keeps moving towards higher values (exemplified on the right in Fig. 4). The WAV problem goes as high as 37 dimensions, and curiously this is one of the problems where M3GP produces substantially smaller trees than M2GP.

The comparison between M3GP and the state-of-the-art classifiers is based only on training and test fitness, once again considering fitness to be accuracy. Based on the comparison previously done between M2GP and several state-of-the-art methods (see Sect. 7.1), we have decided to compare M3GP with a tree based

Table 4 Comparison between the three methods and state-of-the-art classifiers

	HRT	IM-3	WAV	SEG	IM-10	YST	VOW	M-L
Training fitness								
RF	**98.4**	**100**	**99.5**	**99.9**	**99.8**	**98.3**	99.9	99.2
RS	88.9	97.1	92.0	98.4	96.3	71.1	97.8	92.3
MLP	**98.4**	98.7	98.5	97.6	91.0	64.6	91.9	91.3
M2GP	89.4	98.2	87.4	96.8	91.4	62.6	95.9	**100**
M3GP	94.7	99.6	90.7	98.1	93.0	68.5	**100**	**100**
eM3GP	86.7	98.2	81.8	96.1	92.0	61.0	87.8	**100**
Test fitness								
RF	80.2	94.8	81.5	**97.3**	**96.9**	**57.5**	89.4	71.8
RS	**81.5**	92.8	82.2	96.0	93.9	56.6	82.8	65.7
MLP	80.2	**95.9**	83.3	96.3	90.2	**58.0**	82.5	**75.9**
M2GP	80.2	93.8	**84.9**	95.6	90.2	53.8	85.9	63.0
M3GP	79.0	**95.4**	**84.3**	95.6	91.0	56.2	**93.8**	57.1
eM3GP	**81.4**	93.2	81.2	94.7	90.3	56.1	78.6	65.1

The values refer the medians of 30 runs. The best values are in bold (if more than one, it means there is no statistically significant difference between the medians)

classifier (RF—Random Forests), a meta classifier (RS—Random Subspace), and a function based classifier (MLP—Multi Layer Perceptron). The three of them were well ranked in the previous comparison with M2GP. We have also included M2GP in this comparison to check how much better M3GP compares to the state-of-the-art than M2GP. Also eM3GP is included in this table, whose results will be discussed later.

Table 4 reports and compares the training and test fitness obtained by RF, RS, MLP, M2GP and M3GP (and eM3GP, to be discussed later) on the same eight problems, medians of 30 runs. The best approach on each problem is marked in bold, or several when their differences are not statistically significant. Looking at the first row, it is undeniable that RF is an almost unbeatable method when it comes to training fitness. Still, it is beaten by M3GP in the last two problems (VOW and M-L). (M2GP achieves the same feat in only one of them, M-L).

Also in test fitness RF is the best method, ranked first in five of the eight problems. However, other methods are not far from this achievement, including M3GP and MLP, both ranked first in four of the eight problems. (M2GP achieves this is only two problems). While RF is not equaled by any other method in two problems (SEG and IM-10), both M3GP and MLP achieve this in one problem (VOW and M-L, respectively). M2GP and M3GP stand together as winners on the WAV problem.

Besides the remarkable fact that M3GP achieves almost the same quality of results as the popular and successful RF in terms of test fitness, it is also worth remarking that the models provided by M3GP are potentially much easier to interpret than the ones provided by RF or any of the other two state-of-the-art methods.

7.3   Results of eM3GP

The results achieved by eM3GP can be found in Tables 3 and 4. Although eM3GP was not able to match the performance of M3GP in most problems, it appears to be much more resistant to overfitting, based on the observed difference between training and test fitness (Table 3). The obvious case is the M-L problem, where eM3GP actually achieves significantly better test fitness, but also in problems like WAV, SEG and YST it is clear that with eM3GP the test fitness follows the training fitness much more closely than with M3GP. Therefore, even if the final solutions may not necessarily be better with eM3GP, the results suggest that with more generations the tendency may be inverted, and eM3GP may actually be able to reach better performance than M3GP.

One point where eM3GP clearly wins is, no doubt, the compactness of the evolved solutions, both in terms of number of nodes and in terms of number of dimensions (Table 3). The ensemble approach is able to maintain smaller solutions, preventing the bloat at the dimension level and thus performing a kind of dimensionality reduction, at least when compared to M3GP. Also when comparing to M2GP the solutions of eM3GP are much smaller, even in the few cases where the number of dimensions is not.

If we inspect the distribution of dimensions inside the population we can see there is a large difference in its evolutionary dynamics between M3GP and eM3GP. While M3GP tends to produce unimodal distributions that approximate a Gaussian form (see Fig. 4), eM3GP maintains a higher diversity of dimensions within the population, which is either approximately flat (e.g., VOW) or has a single tail with a peak in unidimensional transformations (e.g., M-L), as shown in Fig. 5 (compare the VOW dynamics with the one seen in Fig. 4 for M3GP). Such a distribution, and the effect it has on bloat, seems to correlate nicely with recently proposed bloat control strategies (Silva 2011).

8   Conclusions

This work has addressed the problem of multiclass classification with GP, an area where previous GP approaches tended to yield poor performance. It has presented three variants of a novel method, respectively called M2GP (Ingalalli et al. 2014), M3GP (Muñoz et al. 2015) and eM3GP. The novelty of M2GP is mainly its fitness function, that implicitly drives the evolution into forming multidimensional clusters that allow an accurate classification of the data. M3GP allows the evolution to

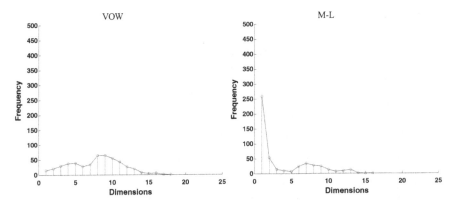

Fig. 5 Distribution of the number of dimensions in the final generation for eM3GP. On the *left*, a typical run of problem VOW. On the *right*, a typical run of problem M-L

choose the most appropriate dimensionality for this mapping during the search, thus improving the adaptability to the particularities and difficulties posed by each problem. Finally, the newer eM3GP introduces an ensemble approach that allows the evolution of specialized mappings for different classes, thus providing some protection against overfitting and the negative effects of class imbalance. As a welcome side effect, eM3GP also removes the bloating problems that seems to affect M3GP.

The results have shown that this new approach finally allows GP to be considered as a viable and competitive option for solving multiclass classification problems, even when compared to the best and most popular state-of-the-art classifiers, like Random Forests, Random Subspaces and Multilayer Perceptron.

Future work will focus on the difficulties of real-world problems. The apparent ability of eM3GP for dealing with overfitting and class imbalance will be thoroughly tested, and certainly improved. We will also go back to the original fitness function of M2GP and improve this core element of success, as it is still in its original "raw" form and its robustness can certainly be improved in order to face the difficulties of real-world data. Another path of future work is the interpretation of the solutions returned by this method. Until now there was absolutely no attempt at performing a symbolic simplification of the mappings returned, or any type of interpretation of what these mappings may reveal about the data.

For now, it is clear that with this new approach we have a simple and general purpose classifier that is well worth testing, improving and using in challenging classification tasks.

Acknowledgements This work was partially supported by FCT funds (Portugal) under contract UID/Multi/04046/2013 and projects PTDC/EEI-CTP/2975/2012 (MaSSGP), PTDC/DTP-FTO/1747/2012 (InteleGen) and EXPL/EMS-SIS/1954/2013 (CancerSys). Funding was also provided by CONACYT (Mexico) Basic Science Research Project No. 178323, DGEST (Mexico) Research Projects No. 5149.13-P and 5414.11-P, and FP7-Marie Curie-IRSES 2013 project ACoBSEC. Finally, the second author is supported by scholarship No. 372126 from CONACYT.

References

Alcala-Fdez J, Fernandez A, Luengo J, Derrac J, Garcia S, Sanchez L, Herrera F (2011) Keel data-mining software tool: data set repository, integration of algorithms and experimental analysis framework. J Mult Valued Logic Soft Comput 17:2–3, 255–287

Bache K, Lichman M (2013) UCI machine learning repository, university of California, Irvine, school of information and computer sciences. http://archiveicsuciedu/ml

Bojarczuk CC, Lopes HS, Freitas AA (2000) Genetic programming for knowledge discovery in chest-pain diagnosis. IEEE Eng Med Biol Mag 19(4):38–44. http://ieeexplore.ieee.org/iel5/51/18543/00853480.pdf

Castelli M, Silva S, Vanneschi L, Cabral A, Vasconcelos MJ, Catarino L, Carreiras JMB (2013) Land cover/land use multiclass classification using gp with geometric semantic operators. In: EvoApplications'13. Springer, Berlin, pp 334–343

Espejo PG, Ventura S, Herrera F (2010) A survey on the application of genetic programming to classification. Trans Sys Man Cyber Part C 40(2):121–144

Falco ID, Cioppa AD, Tarantino E (2002) Discovering interesting classification rules with genetic programming. Appl Soft Comput 1(4):257–269

Haynes T (1998) Collective adaptation: the exchange of coding segments. Evol Comput 6(4):311–338. doi:10.1162/evco.1998.6.4.311. http://dx.doi.org/10.1162/evco.1998.6.4.311

Hsu CW, Lin CJ (2002) A comparison of methods for multi-class support vector machines. IEEE Trans Neural Netw 13(2):415–425

Ingalalli V, Silva S, Castelli M, Vanneschi L (2014) A multi-dimensional genetic programming approach for multi-class classification problems. In: Nicolau M, et al. (eds) 17th European conference on genetic programming. Lecture notes in computer science, vol 8599. Springer, Granada, pp 48–60

Jabeen H, Baig AR (2013) Two-stage learning for multi-class classification using genetic programming. Neurocomputing 116:311–316

Kishore JK, Patnaik L, Mani V, Agrawal VK (2000) Application of genetic programming for multicategory pattern classification. IEEE Trans Evol Comput 4(3):242–258

Koza JR (1992) Genetic programming: volume 1, On the programming of computers by means of natural selection, vol 1. MIT Press, New York

Koza JR (2010) Human-competitive results produced by genetic programming. Genet Program Evolvable Mach 11(3–4):251–284

Langdon W, Poli R (2002) Foundations of genetic programming. Springer, Berlin

Li XM, Wang M, Cui LJ, Huang DM (2007) A new classification arithmetic for multi-image classification in genetic programming. In: International conference on machine learning and cybernetics, vol 3, pp 1683–1687, 2007

Lin JY, Ke HR, Chien BC, Yang WP (2007) Designing a classifier by a layered multi-population genetic programming approach. Pattern Recogn 40(8):2211–2225

Lin JY, Ke HR, Chien BC, Yang WP (2008) Classifier design with feature selection and feature extraction using layered genetic programming. Expert Syst Appl 34(2):1384–1393

Muñoz L, Silva S, Trujillo L (2015) M3gp—multiclass classification with gp. In: Machado P, Heywood MI, McDermott J, Castelli M, García-Sánchez P, Burelli P, Risi S, Sim K (eds) Genetic programming. Lecture notes in computer science, vol 9025. Springer International Publishing, Berlin, pp 78–91

Muni D, Pal N, Das J (2004) A novel approach to design classifiers using genetic programming. IEEE Trans Evol Comput 8(2):183–196

Poli R, Langdon WB, McPhee NF (2008) A field guide to genetic programming. Published via http://lulu.com and freely available at http://www.gp-field-guide.org.uk. http://www.gp-field-guide.org.uk. (With contributions by J. R. Koza)

Sakprasat S, Sinclair M (2007) Classification rule mining for automatic credit approval using genetic programming. In: IEEE congress on evolutionary computation, 2007. CEC 2007, pp 548–555

Shen S, Sandham W, Granat M, Dempsey MF, Patterson J (2003) A new approach to brain tumour diagnosis using fuzzy logic based genetic programming. In: Engineering in medicine and biology society, 2003. Proceedings of the 25th annual international conference of the IEEE (volume 1), vol 1, pp 870–873

Shiming Xiang FN, Zhang C (2008) Learning a mahalanobis distance metric for data clustering and classification. Pattern Recogn 41(2):3600–3612

Silva S (2011) Reassembling operator equalisation: A secret revealed. In: Proceedings of the 13th annual conference on genetic and evolutionary computation, GECCO '11. ACM, New York, pp 1395–1402

Silva S, Tseng YT (2008) Classification of seafloor habitats using genetic programming. In: Applications of evolutionary computing. Lecture notes in computer science, vol 4974. Springer, Berlin, pp 315–324

Tackett WA (1993) Genetic programming for feature discovery and image discrimination. In: Proceedings of the 5th international conference on genetic algorithms. Morgan Kaufmann Publishers Inc, San Francisco, CA, pp 303–311

Tan KC, Tay A, Lee T, Heng CM (2002) Mining multiple comprehensible classification rules using genetic programming. In: Proceedings of the 2002 congress on evolutionary computation. 2002. CEC '02, vol 2, pp 1302–1307

Teredesai A, Govindaraju V (2004) Issues in evolving gp based classifiers for a pattern recognition task. In: Congress on evolutionary computation, 2004. CEC2004, vol 1, pp 509–515

Winkler S, Affenzeller M, Wagner S (2007) Advanced genetic programming based machine learning. J Math Model Algorithm 6(3):455–480

Zhang M, Ciesielski V (1999) Genetic programming for multiple class object detection. In: Advanced topics in artificial intelligence. Lecture notes in computer science, vol 1747. Springer, Berlin, pp 180–192

Zhang M, Smart W (2004) Multiclass object classification using genetic programming. In: Applications of evolutionary computing. Lecture notes in computer science, vol 3005. Springer, Berlin, pp 369–378

Zhang Y, Rockett PI (2009) A generic multi-dimensional feature extraction method using multiobjective genetic programming. Evol Comput 17(1):89–115

Prime-Time: Symbolic Regression Takes Its Place in the Real World

Sean Stijven, Ekaterina Vladislavleva, Arthur Kordon, Lander Willem, and Mark E. Kotanchek

Abstract In this chapter we review a number of real-world applications where symbolic regression was used recently and with great success. Industrial scale symbolic regression armed with the power to select right variables and variable combinations, build robust trustable predictions and guide experimentation has undoubtedly earned its place in industrial process optimization, business forecasting, product design and now complex systems modeling and policy making.

Keywords Symbolic regression • Forecasting • DataModeler • Extrapolation • Prediction • Simulation-based optimization

1 Introduction

Symbolic regression remains the poster child for real-world application of genetic programming and over the past quarter-century has moved from discovering the low-order polynomials of toy data sets to extracting insights, models and profits from data sets ranging up to millions of records and thousands of variables. The ability to simultaneously explore the worth of different variable combinations during

S. Stijven (✉)
Department of Mathematics - Computer Sciences, University of Antwerp, Antwerp, Belgium
e-mail: sean.stijven@uantwerp.be

E. Vladislavleva
Evolved Analytics Europe BVBA, Beerse, Belgium
e-mail: katya@evolved-analytics.be

A. Kordon (retired)
Kordon Consulting LLC, Fort Lauderdale, FL, USA
e-mail: arthur@evolved-analytics.com

L. Willem
Faculty of Medicine and Health Sciences, University of Antwerp, Antwerp, Belgium
e-mail: lander.willem@uantwerp.be

M.E. Kotanchek
Evolved Analytics LLC, Midland, MI, USA
e-mail: mark@evolved-analytics.com

© Springer International Publishing Switzerland 2016
R. Riolo et al. (eds.), *Genetic Programming Theory and Practice XIII*,
Genetic and Evolutionary Computation, DOI 10.1007/978-3-319-34223-8_14

the model development is a huge advantage of multi-objective symbolic regression and provides the foundations for the use cases which are discussed in this chapter.

We open the discussion with a brief review of the role of symbolic regression in process optimization. Although such is one of the earliest real-world applications of GP, success in this space fueled a sizable investment in developing new algorithms, analysis infrastructure and workflow which, subsequently, enabled application in other domains.

The maturation of the capabilities has enabled a migration away from the world of corporate R&D and into the business mainstream. The business forecasting application discussed in the second section features the hybridization and integration of symbolic regression with other technologies—in this case, ARIMAX modeling—which is facilitated by the white-box nature of the developed models.

The final sections are devoted to the development of a metamodel (or a surrogate model) to summarize the propagation of infectious disease for purposes of social policy design. As such it makes the results of highly sophisticated simulation models accessible to policy makers who, via an interactive interface, explore the impact and implications of different assumptions and scenarios. The immediacy of the interaction enables the an awareness and integration of insight which would not be possible if review of myriad charts, tables and text were required.

In summary, over the course of a quarter-century, symbolic regression has moved from the realm of the research lab into manufacturing to business operations and policymaking. This indicates a level of maturity such that it is ready to take its proper place among the pantheon of data analysis tools

2 Modern Process Optimization

2.1 Historical Foundations

One of the first applications of genetic programming in Dow Chemical was in continuous process modeling (Smits and Kotanchek 2004; Kordon and Smits 2001). Over the subsequent 25 years, orders-of-magnitude improvements in both the algorithmic efficiency of symbolic regression as well as corresponding orders-of-magnitude enhancements in compute capabilities has greatly expanded the impact and efficacy of symbolic regression in this realm. In parallel with the model generation capability improvements, new tools and techniques for both designing the modeling strategy and extracting insight from the new plethora of developed models have greatly enhanced the impact. In many respects, the overarching analysis workflow improvement has been as meaningful as the improvement in the model development foundations.

2.2   Corporate Goal

The generic goal of any symbolic regression exercise is to determine an input-output relationship and, furthermore, to determine which of the inputs are most effective and useful in predicting the targeted output. However, there can be nuances which motivate different models being studied than those with the "best accuracy". To illustrate, if new customer requirements are imposing better cold-weather performance for a biofuels additive, we might want to simultaneously understand the chemistry trade-offs involved in achieving the targeted performance while also identifying the operational process control settings and feedstock characteristics required to satisfy the customer.

The net conclusion is that insight and operational performance are both important from the practitioner viewpoint.

2.3   The Importance of Methodology and Workflow

The total-cost-of-ownership of a model is very important since both efficacy and efficiency make or break the ROI of modeling. In a typical industrial analysis, the actual model development tends to take a relatively small fraction of the human time expended. As a result the infrastructure around exploring the available data to design the analysis approach as well as the tools to select models and extract insight are critical ingredients to success.

2.4   Data Exploration and Analysis Design

A conventional assumption in many modeling techniques is to presume that the inputs and independent. This convention is generally violated in most real-world problems as illustrated in Fig. 1 for a biofuel data set. Although variable orthogonality can be achieved, for example, by a principle components analysis, doing such eliminates the interpretability of models beyond the first one or two principle components. Simply pruning the input set to selected sets of independent inputs is not an attractive alternative since such imposes a priori constraints which are not necessary and are often counter-productive.

2.5   Model Development

The basics of the evolutionary search for symbolic regression models are pretty straight-forward: reward models for accuracy, simplicity and novelty [Smits and Vladislavleva; Vladislavleva et al.] and let the primordial soup percolate. The art

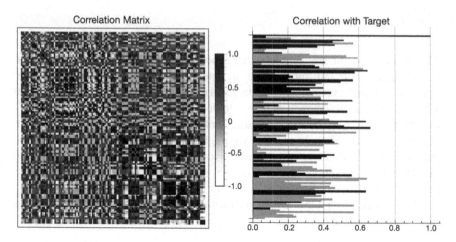

Fig. 1 The *left plot* illustrates the correlation matrix for the available inputs for a biofuels analysis. Many of the available process setpoints and measurements as well as the associated chemical analysis results in the data are highly correlated (either positively or negatively) which implies that we will have potential for variable substitutions in the model development. Note in this case that the maximum correlation is less than 0.7 which implies a foundation capability of $R2 < 0.5$ for a simplistic single-variable model

comes in the selection of available function operators, definition of metavariables and normalization of data ranges. From a user perspective, they care mostly about having a selection of high-quality models which properly balance the complexity-accuracy trade-off since our goal is to have as simple of model as possible—but no simpler. Such a collection of models is illustrated in Fig. 2.

2.6 Model Exploration and Insight Development

To paraphrase Bill Worzel, "Symbolic regression is an optionizer as well as optimizer." This is illustrated in Figs. 3 and 4. As such many potential models are hypothesized, refined (or rejected) and available at the end of the development stage. Since each of the independent model searches follows its own path through the search space, we have the raw material to collect considerable insight into the modeling potential and alternate solutions. An analysis often requires multiple rounds of model development with each iteration building upon the insights gathered from the prior rounds to focus the input variable set, tune development options or simply to evolve additional model forms.

Understanding the number of variables required to achieve a given level of performance and the modeling potential of inputs is very important. For instance, even though a particular input does not quite provide as direct of path to a quality model as another, it may be a measurement that is more easily or robustly achieved which, from an operational standpoint, would make models containing it rather than it the more desirable.

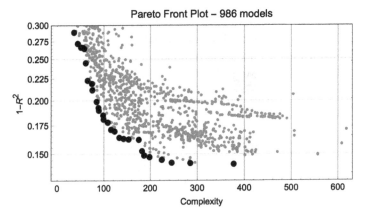

Fig. 2 Although additional attributes may be used as criteria in the model development, from a practical standpoint, the candidate model set comes from those models which best balance the complexity-accuracy trade-off. In this figure, the models denoted as *red dots* are those lying on the Pareto front. Although these are nominally optimal, the other models indicated in *blue* in the density plot may be of more practical interest due to their model dimensionality or particular combination of constituent models (Color figure online)

2.7 Model Selection and Ensemble Definition

The deterministic nature of the physical sciences introduces a bias towards THE model. However, the stochastic nature of evolutionary search will uncover many "good enough" models which we can judiciously combine to create a trustable model that will warn if it is exposed to new operating conditions or if the modeled system has undergone some sort of fundamental change so that the models are no longer applicable. This can be easily accomplished by selecting models from near the knee of the Pareto front which have uncorrelated error residuals. Because the selected ensemble models agree with the development data, they will agree when near known operating conditions and, because they are diverse, they will diverge when exposed to new operating regions. This is illustrated in Fig. 6 for the model set shown in Fig. 5.

2.8 Process Optimization Summary

The computational and algorithmic advances over the past two decades allows symbolic regression to quickly build insightful models from process data. These can then be deployed as inferential sensors to control and optimize the targeted processes as well as to provide guidance for operational opportunities and enhancements.

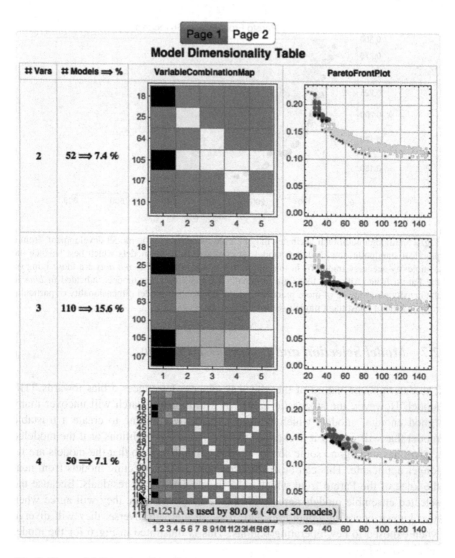

Fig. 3 The model dimensionality table shows the modeling potential achieved as a function of the number of inputs in the constituent models. In this biofuel distillation case, it also illustrates that there are many paths to comparable models due to potential variable substitutions

3 Modern Business Forecasting

Another new area of industrial applications of GP is business forecasting. The ultimate objective of business forecasting is to deliver to the key decision-makers a reliable forecast on specific economic variables, such as product demand, raw materials prices, labor cost, etc. Of special interest are the forecasting methods

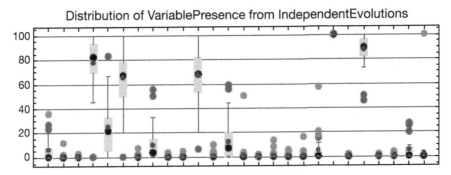

Fig. 4 The founders effect means that each model development follows its own path through search space. Here we look at the variables which have emerged to the fore. Because of the couplings of inputs, we can see that some inputs are clearly desirable. Others can clearly be ignored and some may be substituted for others and, in some circumstances or combinations, become useful

"Model Selection Report"

	Complexity	$1-R^2$	Function
1	27	0.028	$0.72 - \left(4.95 \times 10^{-3}\right) \text{refluxFlow} - \frac{392.81}{\text{vaporFlow}} + \left(7.86 \times 10^{-2}\right) \text{temperatureTray}_1$
2	29	0.021	$-0.13 + \frac{\left(7.59 \times 10^{-8}\right) \text{vaporFlow}^2 \, \text{temperatureTray}_1{}^3}{\text{refluxFlow}}$
3	33	0.030	$-3.60 + 0.43 \left(-\text{refluxFlow} + \text{vaporFlow}\right)^{1/3} + \left(7.86 \times 10^{-2}\right) \text{temperatureTray}_1$
4	34	0.016	$-0.13 + \frac{\left(1.07 \times 10^{-5}\right) \text{vaporFlow}^2 \, \text{temperatureTray}_1{}^3}{\text{refluxFlow}^2}$
5	36	0.023	$4.13 \times 10^{-2} - \left(7.97 \times 10^{-7}\right) \left(-\text{refluxFlow} + \text{vaporFlow}\right) \left(9 - \text{temperatureTray}_1\right)^3$
6	39	0.021	$-0.13 - \frac{\left(7.17 \times 10^{-8}\right) \text{vaporFlow}^2 \, \text{temperatureTray}_1{}^3}{7.87 - \text{refluxFlow}}$
7	39	0.023	$0.63 - \left(1.53 \times 10^{-2}\right) \text{refluxFlow} + \left(6.92 \times 10^{-3}\right) \text{vaporFlow} - \left(3.17 \times 10^{-2}\right) \text{temperatureTray}_1 + \left(1.58 \times 10^{-3}\right) \text{temperatureTray}_1{}^2$
8	43	0.016	$-2.55 - \left(1.17 \times 10^{-2}\right) \text{refluxFlow} - \frac{7965219.60}{\text{vaporFlow}^3} + \frac{44.29}{\text{temperatureTray}_1} + 0.13 \, \text{temperatureTray}_1$
9	43	0.020	$4.89 - \left(1.58 \times 10^{-2}\right) \text{refluxFlow} + 0.87 \, \text{vaporFlow}^{1/3} - 4.61 \, \text{temperatureTray}_1{}^{1/3} + 0.23 \, \text{temperatureTray}_1$
10	47	0.016	$-17.34 - \left(1.29 \times 10^{-2}\right) \text{refluxFlow} + 1.62 \sqrt{\text{vaporFlow}} - \left(4.45 \times 10^{-2}\right) \text{vaporFlow} + \frac{44.00}{\text{temperatureTray}_1} + 0.13 \, \text{temperatureTray}_1$
11	55	0.013	$-\left(7.41 \times 10^{-2}\right) - \left(2.15 \times 10^{-3}\right) \text{temperatureTray}_1{}^{3/2} + \frac{\left(1.05 \times 10^{-7}\right) \text{vaporFlow}^2 \, \text{temperatureTray}_1{}^3}{\text{refluxFlow}}$
12	63	0.013	$-0.32 + \left(2.39 \times 10^{-2}\right) \text{temperatureTray}_1 - \left(4.81 \times 10^{-3}\right) \text{temperatureTray}_1{}^{3/2} + \frac{\left(1.03 \times 10^{-7}\right) \text{vaporFlow}^2 \, \text{temperatureTray}_1{}^3}{\text{refluxFlow}}$
13	66	0.010	$-18.58 - \frac{1377.77}{\text{refluxFlow}} - \left(3.30 \times 10^{-2}\right) \text{refluxFlow} + 9.03 \, \text{vaporFlow}^{1/3} - \left(6.84 \times 10^{-2}\right) \text{vaporFlow} - 4.29 \, \text{temperatureTray}_1{}^{1/3} + \frac{35.05 \, \text{temperatureTray}_1}{\text{refluxFlow}}$
14	66	0.014	$0.41 - \left(1.82 \times 10^{-2}\right) \text{temperatureTray}_1 - \left(1.59 \times 10^{-6}\right) \text{vaporFlow} \, \text{temperatureTray}_1{}^{3/2} + \frac{\left(1.33 \times 10^{-7}\right) \text{vaporFlow}^2 \, \text{temperatureTray}_1{}^3}{\text{refluxFlow}}$

Fig. 5 This is a table of diverse 3-variable ensemble models for a distillation column

based on explanatory variables (economic drivers), the most popular of which is the Auto-Regressive Integrated Moving-Average with eXplanatory variables (ARIMAX) model. A limitation of this approach, however, is the assumption of linear relationships between the explanatory variables and the target variable.

One option to overcome this limitation is by integrating GP and ARIMAX methods into a hybrid system where the nonlinearity is represented with transforms, generated by GP and used as explanatory variables in ARIMAX models.

The final forecast is generated by ARIMAX models with nonlinear explanatory variables (Kordon 2014). The nonlinear equations, generated by GP, can be based

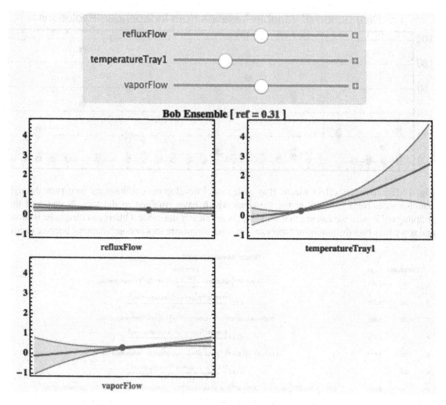

Fig. 6 The constituent models of an ensemble will diverge when asked to extrapolate. This is illustrated for a 3-variable model ensemble consisting of 14 unique models

on both contemporaneous and dynamic relationships. The last option assumes using lagged inputs up to an expected maximum lag. The selected nonlinear transforms from the GP-generation phase are used in ARIMAX model generation but they might not be selected in the final model if they are statistically insignificant. If distributed lags of a statistically significant nonlinear transform exist, it is represented by the corresponding transfer function in the ARIMAX model. As a result, GP complements ARIMAX models with adding contemporaneous and dynamic nonlinear explanatory variables and the final models are with all the benefits of this well-known forecasting approach, such as building multi-step forecasts of all inputs, statistically defined confidence limits, and available software.

An important area of business forecasting with big economic impact is raw materials prices forecasting where 3-to-6 months forecasts are critical in high-volume price negotiation. A recent application of large-scale raw materials forecasting in the chemical industry is discussed in (Kordon 2012). An example of applying the hybrid system in two typical cases in raw materials forecasting: (1) when the relationships between the forecasted variable and the related economic drivers are contemporaneous and (2) when the relationships are dynamic due to lags, is shown in Kordon (2014). In the first case, a simple nonlinear transform, generated by GP and used as exogenous input in the ARIMAX model, has shown the best ex-ante performance. In the second case, GP has generated a dynamic model with accurate lags.

This model has been compared to another contemporaneous nonlinear model, generated by GP, and the best linear ARIMAX model. The ex-ante performance of the dynamic model is the best for the tested period of time. These encouraging results based on real world applications with high economic impact demonstrate the big potential of GP in business forecasting.

4 Modern Complex Systems Analysis and Policy Making

For many years already we have been proposing to use symbolic regression for simulation-based optimization, with its claim to fame being an white-box model which can be built to model the behaviors of a complex simulator and a built-in variable selection capability, which will only be producing models using variables impacting simulator responses. The more robust are the symbolic regression features—the more complicated simulators we can consider to understand and meta-model. The process of building models of simulation models is referred to as meta-modeling, and the process of optimizing and understanding the models (as well as the underlying simulators and complex systems they are mimicking) is referred to as simulation-based optimization, derivative-free optimization.

In this section we share the results of an exciting project to understand an truly complex system—the state of the art simulator of the spread of infectious diseases in a large population of humans, based on probabilistic individual-based models. This inter-disciplinary project took place in Belgium by several groups and led to a publication on Active Learning to Understand Infectious Disease Models and Improve Policy Making at PLOS Computational Biology by Lander Willem, Sean Stijven, Ekaterina Vladislavleva, Jan Broeckhove, Philippe Beutels, and Niel Hens.

Results presented in this section appeared in Active Learning to Understand Infectious Disease Models and Improve Policy Making at PLOS Computational Biology by Lander Willem, Sean Stijven, Ekaterina Vladislavleva, Jan Broeckhove, Philippe Beutels, and Niel Hens (DOI: 10.1371/journal.pcbi.1003563) and are presented here for the audience interested in complex systems analysis and simulation-based optimization.

By large population we mean demographies of 0.5–300 million people in known geographies (cities, counties, countries) and known probabilistic contact networks. Population is decided into age groups from the known distribution, into families, "children" go to school and kindergarten, adults travel to work by cars and public transportation, communicate with their colleagues at work during the day and with their families in the evening. In the presence of an infections disease with a certain infectious rate R0, each individual can be in one of the four states—healthy, infected and being infectious without symptoms, infectious while displaying symptoms, and recovered.

The basic reproduction number R_0 is the most important indicator of how dangerous the infectious disease is. R_0 is defined as the expected number of secondary infections caused by a primary infection in a fully susceptible population. In the presence of an actual pandemics threat, R0 can be estimated from a small population of early infected cases quite accurately within a couple of days. For the NHN1 influenza virus the R0 value observed was 1.8.

Robust evaluation of worst and best case scenarios, realistic predictions of whether or not the pandemics will happen and robust forecasts of the damages, are of critical importance to the governments and National Health institutes. This global economic impact and the need to understand pandemics has made influenza the subject of many simulation studies. The most recent state-of-the art dynamic models for influenza are a lot more realistic than static compartmental deterministic models, but realism comes at price of much higher complexity. Dynamic individual-based models require a lot more simulation time, provide less transparency and are looked at as a black-box simulators with many "knobs to turn" of unknown significance for simulation results.

Many fundamental parameters in individual-based models are unknown a priori and the computational complexity severely hinders experimental approaches with different scenarios of distributions of these fundamental parameters.

Computational complexity and the presence of too many variables of unknown significance as well as the need to understand relationships and dependencies in input response data make symbolic regression the perfect tool for model-based understanding of dynamic models.

In Willem et al. (2014) the authors proposed a hybrid approach for understanding the complex black-box simulations through an iterative strategy of collecting the simulation data, reducing dimensionality, identifying new optimal experiments and repeating the process until ensembles of reliable and transparent symbolic regression meta-models is found and can be used to sensitivity analysis and design space exploration.

4.1 Flute Simulator

As a basis for a simulator we took an open-source individual-based model for influenza epidemics called FluTE and developed by Chao et al. (2010). This state of the art model is written in C++. All individuals are simulated as members of different social mixing groups (Fig. 7). Within each group influenza transmission is based on random mixing. The geographical distribution, employment rates and commuting behavior of the population are based on the 2000 Census data for Seattle (500,000 people) and the Los Angeles County (11 million people). The data is distributed together with the source code of the model. The simulation runs for 180 days in 12-h time steps, representing daytime (work, school and community contacts) and nighttime (home and community contacts). The contact probabilities in the model were tuned such that the final age-specific clinical attack rates were

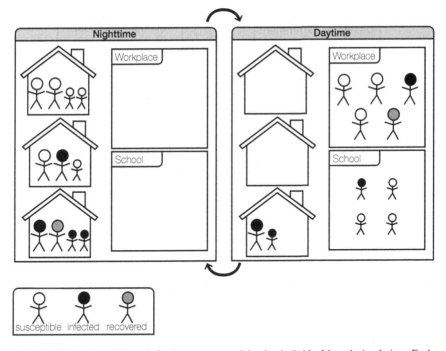

Fig. 7 An illustration of how infections are spread in the individual-based simulation. Each simulation run models 180 days in 12-h intervals (daytime and nighttime), and at the end the resulting outcomes are calculated,—the total attack rate (the fraction of the population that got infected) and the peak day (the day of the 180-day interval with the maximum attack rate)

similar to past influenza pandemics and observed household attack rates.[1] This FluTE model can simulate several intervention strategies like vaccination or spread of antiviral medications with subsequent changes in susceptibility and infectivity as well as social distancing measures like quarantines and subsequent changes in contact probabilities between individuals.

4.2 Experimental Workflow

We provide an overview of the experimental workflow in Fig. 8. This section follows the order of the steps as depicted in Fig. 8.

---

[1] Attack rate is defined as a ratio of the new cases in the population at risk to the total size of the population at risk.

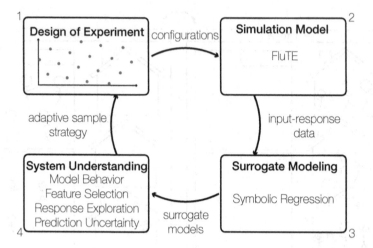

Fig. 8 (Adapted from Willem et al. 2014) Workflow of the data-driven simulator analysis and understanding. The goal of such a workflow is to understand which inputs to the simulator are impacting the outputs, and build robust and efficiently evaluation able meta-models to reliably mimic the simulator behavior

4.2.1   Step 1: Design of Experiments

To maximize the information content from the experiments in the high-dimensional input space with inputs of unknown significance we used space-filling Latin hypercube designs (LHD) and their approximations to create input data sets for the FluTE simulations. Latin Hyper cube designs are particularly useful to prevent the "collapse" in the input data design in cases where input variables might turn out to be insignificant and will be omitted from consideration. Space-filling designs are critical in simulation-based optimization applications, where the variable selection and system understanding happens iteratively and data collection is precious.

In the general case, a sample value from the first interval of the first input parameter is matched at random with sample values from intervals chosen for the other input parameters (Ma et al. 1993). Then the second interval of the first input parameter is matched at random with sample values from previously unused intervals of the other features. Each interval of every input parameter will be sampled once and only once.

LHD has the advantage that the number of samples is independent of the number of dimensions of the input space but can be determined based on the computational budget, the input dimensions and the complexity of the simulation. Computing a space-filling LHD can be an onerous task and therefore we used the maximin designs of spacefillingdesigns.nl (Santner et al. 2003; Husslage et al. 2006). Because our designs have a rather limited number of sample points, we extended the designs using the Intersite-projected distance method of the Sequential Experimental Design (SED) toolbox (Crombecq and Dhaene 2010; Crombecq et al. 2009; Crombecq 2011).

4.2.2 Step 2: Simulation Model

The FluTE simulator allows varying 38 individual inputs (indicators defining the influenza in question, the number of seed cases (infected people entering the population on day 1), and prevention measures, like closing schools and kindergartens, enforced quarantines, vaccinations (with vaccination fractions and efficacy), antiviral medication and their influence on infection probabilities, etc.

We will add information on how the input-space is converted to simulation configurations. We will add some info on how we executed the simulation on a cluster, and will add info on the avg. calculation time of each simulation run—2 h.

The summary is that with 37 knots to turn and 2 h needed to evaluate one input-response combination (simulating a couple of million people changing states over 180 days and nights), quick data collection is virtually impossible. Meta-modeling of such computationally expensive simulators and development of interactive tools to efficiently explore what-of scenarios is the only solution to prepare for crisis situations. When meta-models are created (with symbolic regression in this case), they can be evaluated in real-time immediately as estimations for the transmission rate R_0 of the attacking infection are becoming available.

4.2.3 Step 3: Meta-Modeling with Symbolic Regression

After the simulation runs are completed, we gather the results and create an input-response data set. Symbolic Regression (SR) applied to this data set will generate a robust ensemble of meta-models emulating the behavior of the simulator. We used the SR implementation from the DataModeler package in Mathematica (Evolved Analytics LLC 2011). The result of a single SR experiment is an ensemble of tree-based regression models that give a good approximation of the response variable together with the confidence metric for the prediction.

In Willem et al. (2014) we used fixed time budgets for SR experiments based on the size and dimensionality of the data sets. Timings are listed in Table 1.

4.2.4 Step 4: System Understanding

To arrive at a convincing ensemble of symbolic regression models we selected ensembles at the "knee" of the Pareto Front of non-dominated trade-offs in model complexity and model error spaces. At the last step the non-linear optimization of constants in all ensemble models was performed. A model ensemble of high-quality and minimal complexity obtained through an effective SR algorithm can facilitate system understanding and focus the research.

The main differentiating factors of an effective SR implementations are the facts that the final ensembles only contain variable drivers, ensembles are constructed from maximally different strong SR learners, and therefore provide reliable confidence intervals for their prediction. Besides variable importance, good final

Table 1 Symbolic regression settings

Name	Value
Population size	1000
Archive size	100
Crossover rate	0.9
Mutation rate	0.1
Population tournaments	5
Primitive functions	$+, -, *, \div, \bullet^{-1}, \bullet^{2}, \bullet^{x}, \sqrt{}, log, exp$
Time budget FluTE RUN 1,2,3	1000 s
Time budget FluTE RUN 4	7200 s
Time budget FluTE RUN 5	3600 s
Independent evolutions FluTE	8

ensembles also provide dimensionality trade-offs in complexity and accuracy of models. Another strong benefit of effective symbolic regression implementation is functionality to automatically generate hypotheses for meta-variables,—low order transformations of driver inputs, which can potentially linearize the final models and enable further application of the powerful linear and regularized linear learning.

The ultimate highlight of SR-enabled system understanding is interactive sensitivity analysis of generated ensembles. Interactive exploration as well as mathematical optimization of SR ensembles allows to identify "edge-cases", which might have been over-looked or un-anticipated by the domain experts. In addition, interactive prediction explorers are the only way to present the solutions and what-if scenario exploration to business decision makers (without overburdening them with mathematical models). Figure 9 illustrates a snapshot of a six-variable prediction explorer for the clinical attack rate. This and other explorer are publicly available at http://www.idm.uantwerpen.be.

4.3 Results

4.3.1 Transmission

As stated above we performed a stepwise exploration of the US-tailored simulation model for pandemic influenza (FluTE), applied to Seattle and Los Angeles county (Chao et al. 2010). We first simulated epidemics in the Seattle population using four basic model parameters: R_0, whether individuals can travel, the number of infected individuals introduced into the population and whether this seeding occurs only once (static) or on a daily basis (dynamic). Table 2 summarizes the parameter ranges.

The surrogate models for the cumulative clinical attack rate(AR) were of good quality (error < 0.001). The cumulative clinical attack rate is the fraction of the population that got infected. Although each configuration was executed 20 times, almost no stochastic fadeout was observed. The dichotomous variable indicating whether people can travel was absent in most surrogate models. Given the inherent

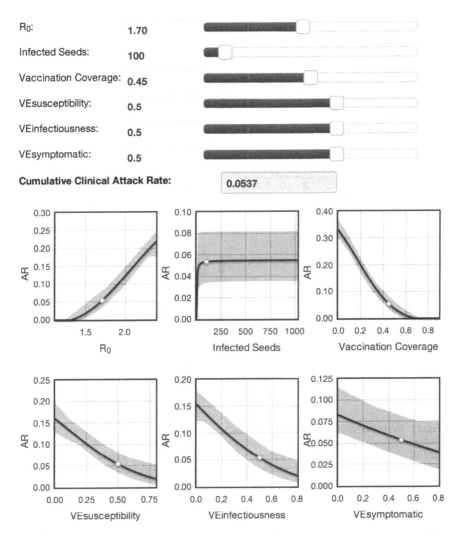

Fig. 9 Response plot explorer for the cumulative clinical attack rate. An interactive version of this plot is available at *www.idm.uantwerpen.be*

feature selection of SR, this parameter appears to be unimportant to predict the AR (Stijven et al. 2011). The response plot for the AR (Fig. 10a) shows that the number of infected people seeded into the population had almost no impact when seeding once. Only very low numbers of seeded individuals resulted in a different AR. The impact of the seeding number on the AR increased with daily seeding. We observed a correlation of 60 % between the AR and the seeding number and frequency though we expected a major role for R_0.

The day of the epidemic peak advanced logarithmically with an increasing number of infected seeds, although small numbers of seeds could give rise to no

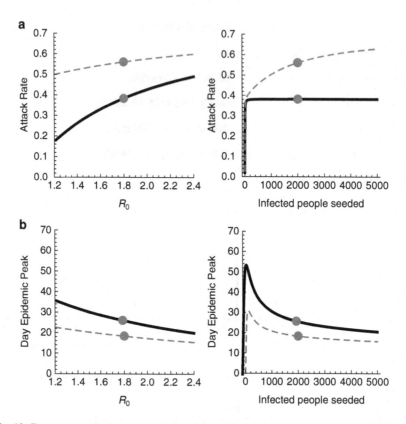

Fig. 10 Response prediction plots of the high-quality surrogate models obtained with SR. [Adapted from Willem et al. (2014)] Response prediction plot for the cumulative clinical attack rate (**a**) and the day on which the epidemic reaches its peak (**b**) when seeding occurs only once (*black*) or on a daily basis (*gray*). Predictions for R_0 assume a fixed number of infected seeds, indicated by the dot in the panel on the right, and vice versa

or very late peaks (Fig. 10b). There is no consensus in the literature on pandemic influenza models about how and to which extent infectious individuals should be seeded. Some studies (Ferguson et al. 2005; Chao et al. 2010; Halloran et al. 2008; Andradóttir et al. 2011) have been published with static seeding of 1, 10 and 100 individuals while others used dynamic seeding. There seems to be no concern about the potential impact of these different seeding approaches, as only a shift of the epidemic curve due to seeding has been reported (Germann et al. 2006). We explored a wide range of seeding values using both static and dynamic approaches, and observed that the seeding approach has impact on the results. The surrogate model divergence for small seeding values was very large so these conditions needed to be sampled more intensively.

Population

In order to assess the effect of population size, we compared FluTE simulations for Seattle (0.5 million people) and LA County (11 million people). We used a single design with four transmission parameters for both populations (Table 2) and compared the surrogate models of each dataset. We observed similar response predictions for the AR (Fig. 11a), indicating that this outcome is insensitive to population size, when population size is already substantial (i.e. 0.5 m). The travel parameter was absent in most surrogate models for both populations, indicating that this is inherent to the simulation model. The main difference for the enlarged population was the timing of the epidemic (Fig. 11b). For example, a pandemic with $R_0 = 1.8$ and 100 infected seeds would result in an AR of 0.38 for both populations, but the epidemic peak day in LA County is predicted to be 15 days later compared to Seattle. The similar AR and postponed peak for the larger population are in line with results of previous studies (Ferguson et al. 2006; Chao et al. 2010). We did not compare urban and rural regions due to lack of data although this may have a large impact (Ferguson et al. 2005). Model ensemble divergence for low seeding numbers was less for LA County, which suggests that large populations absorb stochastic effects.

Vaccination

After adjusting the transmission settings, seven parameters for reactive vaccination strategies were added to the design (Table 2). The computational burden to simulate Seattle was much lower compared to the LA County. Therefore, we used the Seattle population for the initial exploration with vaccination parameters. Based on the resulting input-response data, surrogate modeling showed that mainly the response threshold and ascertainment fraction were important to predict the AR. The importance of R_0 and the vaccination coverage increased when the response threshold and ascertainment parameters were set to mimic instant reactive measures, immediately after emergence.

Emulation

After subsequent simulation and modeling iterations, we obtained surrogate models for LA County that can be used to explore reactive vaccination policies on the outcome of ongoing pandemics. Figure 9 shows a basic interface to visualize the response behavior by changing the surrogate model parameters. When vaccination coverage is set to zero, the results from the second design emerge again (Fig. 10). Further exploration of the surrogate models revealed a saturation effect of the vaccination coverage on the AR. The predicted AR with a vaccination coverage of 60 % is almost zero for $R_0 = 1.8$ and vaccine efficacies of 0.5. The protection of the general population by vaccination of a subset is known as herd immunity (Piedra

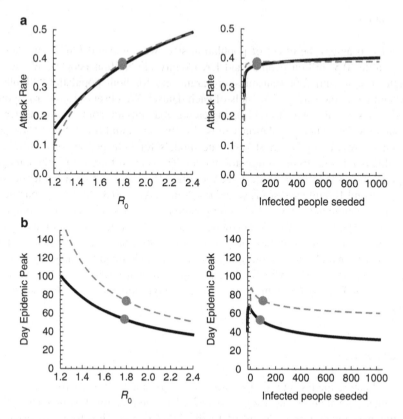

Fig. 11 Response prediction plots for Seattle and LA County. [Adapted from Willem et al. (2014)] Response prediction plots of the high-quality surrogate models obtained with SR for the cumulative clinical attack rate (**a**) and the day of the epidemic peak (**b**) in Seattle (*black*) or LA county (*gray*). Predictions for R_0 assume a fixed number of infected seeds, indicated by the *dot* in the panel on the *right*, and vice versa

et al. 2005). The clear visualization of herd immunity with the surrogate models emphasizes the usefulness of our approach since it is hard to observe this effect directly from the numerous individual simulation results. An interactive version of this plot is available at *www.idm.uantwerpen.be*.

4.4 Summary of Modern Complex System Analysis

Symbolic Regression combined with intelligent design of experiments has proven to be an indispensable tool in understanding complicated simulation models and reducing them to practice. The process of replacing a black-box simulator by an

Table 2 Parameter design for all modeling iterations with FluTE and obtained variable importance for the cumulative clinical attack rate

Parameter	RUN 1	RUN 2	RUN 3	RUN 4	RUN 5
Region	Seattle	Seattle	LA County	Seattle	LA County
Travel allowed?	Yes/no (−)	Yes/no (−)	Yes/no (−)	yes	yes/no (−)
R_0	1.1–2.4 (++)	1.1–2.4 (++)	1.1–2.4 (++)	1.1–2.4 (++)	1.1–2.4 (++)
Infected seeds	0–5000 (+*)	0–1024 (+*)	0–1024 (+*)	0–1024 (+*)	0–1024 (+*)
Seeded daily?	Yes/no (++)	Yes/no (+)	No	Yes/no (-)	No
Ascertainment				0–90 % (+)	80 %
Ascertainment delay				1–5d (−)	1d
Response threshold				0–5 % (+)	instant
Response delay				0–30d (−)	instant
Vaccination coverage				0–90 % (−)	0–90% (++)
$VE_{susceptibility}$				0–66 % (−)	0–66 % (+)
$VE_{infectiousness}$				0–66 % (−)	0–66 % (−)
$VE_{symptoms}$				0–66 % (−)	0–66 % (−)
Scenarios	200	200	50	800	200
Repetitions	20	20	10	20	20

Legend: ++ very important, + important, − almost no impact, * only small values, VE: Vaccine efficacy

interactive profiler of a simulator meta-model provides unique selling points for several reasons:

- It is critical to understand simulator inputs that impact the Key Performance Indicators, and SR does just that.
- It is critical to get reliable predictions for predictions of simulator outputs as well and the trustability of the predictions. Robust ensemble-based symbolic regression does just that.
- The only way to facilitate data-driven decisions is to empower the decision maker with simple to use tools to explore what-of scenarios and be best prepared for whatever is coming.

References

Andradóttir S, Chiu W, Goldsman D, Lee M, Tsui K, Sander B, Fisman D, Nizam A (2011) Reactive strategies for containing developing outbreaks of pandemic influenza. BMC Public Health 11(Suppl 1):S1

Chao D, Halloran M, Obenchain V, Longini I (2010) FluTE, a publicly available stochastic influenza epidemic simulation model. PLoS Comput Biol 6(1):e1000,656

Crombecq K (2011) Surrogate modelling of computer experiments with sequential experimental design. Ph.D. thesis, University of Antwerp, Antwerp

Crombecq K, Dhaene T (2010) Generating sequential space-filling designs using genetic algorithms and monte carlo methods. In: Simulated evolution and learning. Lecture notes in computer science, vol 6457. Springer, Berlin, pp 80–84

Crombecq K, De Tommasi L, Gorissen D, Dhaene T (2009) A novel sequential design strategy for global surrogate modeling. In: Winter simulation conference, Austin, Texas, WSC '09, pp 731–742

Evolved Analytics LLC (2011) DataModeler Release 8.0 Documentation. Evolved Analytics LLC - www.evolved-analytics.com

Ferguson N, Cummings D, Cauchemez S, Fraser C, Riley S, Meeyai A, Iamsirithaworn S, Burke D (2005) Strategies for containing an emerging influenza pandemic in Southeast Asia. Nature 437(7056):209–214

Ferguson N, Cummings D, Fraser C, Cajka J, Cooley P, Burke D (2006) Strategies for mitigating an influenza pandemic. Nature 442(7101):448–452

Germann T, Kadau K, Longini Jr I, Macken C (2006) Mitigation strategies for pandemic influenza in the United States. PNAS 103(15):5935–5940

Halloran M, Ferguson N, Eubank S, Longini I, Cummings D, Lewis B, Xu S, Fraser C, Vullikanti A, Germann T, et al (2008) Modeling targeted layered containment of an influenza pandemic in the United States. PNAS 105(12):4639–4644

Husslage B, Rennen G, Van Dam ER, Den Hertog D (2006) Space-filling Latin hypercube designs for computer experiments. Tilburg University

Kordon AK, Smits GF (2001) Soft sensor development using genetic programming. In: Spector L, Goodman ED, Wu A, Langdon WB, Voigt HM, Gen M, Sen S, Dorigo M, Pezeshk S, Garzon MH, Burke E (eds) Proceedings of the genetic and evolutionary computation conference (GECCO-2001), Morgan Kaufmann, San Francisco, California, pp 1346–1351. http://www.cs.bham.ac.uk/~wbl/biblio/gecco2001/d24.pdf

Kordon AK (2012) Applying intelligent systems in industry: a realistic overview. In proceedings of the 6th IEEE international conference intelligent systems. http://ieeexplore.ieee.org/xpl/articleDetails.jsp?tp=&arnumber=6335108

Kordon AK (2014) Applying genetic programming in business forecasting. Genetic programming theory and practice XI. http://link.springer.com/chapter/10.1007/978-1-4939-0375-7_6

Ma J, Ackerman E, Yang J (1993) Parameter sensitivity of a model or viral epidemics simulated with Monte Carlo techniques. I. illness attack rates. Int J Biomed Comput 32:237–253

Piedra P, Gaglani M, Kozinetz C, Herschler G, Riggs M, Griffith M, Fewlass C, Watts M, Hessel C, Cordova J, et al (2005) Herd immunity in adults against influenza-related illnesses with use of the trivalent-live attenuated influenza vaccine (CAIV-T) in children. Vaccine 23(13):1540–1548

Santner TJ, Williams BJ, Notz WI (2003) The design and analysis of computer experiments. Springer, New York

Smits G, Kotanchek M (2004) Pareto-front exploitation in symbolic regression, Chap. 17 In: O'Reilly UM, Yu T, Riolo RL, Worzel B (eds) Genetic programming theory and practice II. Springer, Ann Arbor, pp 283–299. doi:10.1007/0-387-23254-0_17

Smits G, Vladislavleva E (2008) Trustable symbolic regression models: using ensembles interval arithmetic and pareto fronts to develop robust and trust aware models. In: Dow Benelux BV, Terneuzen (eds) Tilburg University, Tilburg, the Netherlands. Evolved-Analytics, LLC, Midland, MI, USA http://link.springer.com/chapter/10.1007%2F978-0-387-76308-8_12

Stijven S, Minnebo W, Vladislavleva K (2011) Separating the wheat from the chaff: on feature selection and feature importance in regression random forests and symbolic regression. In: Proceedings of the 13th annual conference companion on genetic and evolutionary computation, Dublin, GECCO '11, pp 623–630

Trustable symbolic regression models: using ensembles, interval arithmetic and pareto fronts to develop robust and trust-aware models

Vladislavleva E , Smits G, Kotanchek M (2008) Better solutions faster: soft evolution of robust regression models in pareto genetic programming. In: Dow Benelux BV, Terneuzen (eds) Tilburg University, Tilburg, the Netherlands. Evolved-Analytics, LLC, Midland, MI, USA http://link.springer.com/chapter/10.1007%2F978-0-387-76308-8_2

Willem L, Stijven S, Vladislavleva E, Broeckhove J, Beutels P, Hens N (2014) Active learning to understand infectious disease models and improve policy making. PLoS Comput Biol 10(4). doi:10.1371/journal.pcbi.1003563. http://dx.doi.org/10.1371/journal.pcbi.1003563

Index

A
A*, 34
Affenzeller Michael, 1
against-replication, 76
ancestry, 185, 187, 191, 195
answers, 67

B
Banzhaf Wolfgang, 39
Burlacu Bogdan, 1

C
Castelli Mauro, 219
caveats, 62
choosing-a-representation-language, 70
Classification, 39
classification, multiple classes, clustering, 219
Conserved Stems, 21
Conserved Stems Extraction, 28
Cross-Validation, 79

D
Data Science, 117
DataModeler, 241
Decision-tree, 39
diversity, 117, 152
Donatucci David, 185

E
Elyasaf Achiya, 22
ensembles, 117
Evolutionary Computation, 79

F
Forecasting, 241

G
genealogy, 185, 187, 191, 195
Genetic Algorithms, 79
Genetic Programming, 39, 79
gp-as-if-you-meant-it, 64
gp-as-mangle-ish-practice, 75
gradient boosted regression, 117
graph database, 185, 189
Gustafson Steven, 117

H
Helmuth Thomas, 151, 185
Heuristic Search, 23
Heuristics, 23
HH-Evolver, 24
Hodjat Babak, 79
Hyper Heuristic, 24

I
industrial application, 117
Ingalalli Vijay, 219
initial-setup-and-restrictions, 71

K
Kaizen Programming, 39
Kommenda Michael, 1
Kordon Arthur, xiv, 241
Korns Michael, 203

© Springer International Publishing Switzerland 2016
R. Riolo et al. (eds.), *Genetic Programming Theory and Practice XIII*,
Genetic and Evolutionary Computation, DOI 10.1007/978-3-319-34223-8

Kotanchek Mark, xiv, 241
Krawiec Krzysztof, 170
Kronberger Gabriel, 1

L
lexicase selection, 185, 190, 191
Lexicase selection, diversity, tournament
        selection, implicit fitness sharing, 151
lexicase-selection, 68

M
machine learning, 117
Maximum Weighted Clique, 29
McPhee Nicholas Freitag, 151, 185
Milo Nimrod, 22
Muñoz Luis, 219
multidimensional, balanced accuracy,
        ensemble, 219

N
Narasimhan Ram, 117
Neo4j, 185, 189

O
O'Reilly Una-May, 170
on-mindful-exercises, 61
on-the-mangle-of-practice, 73
operators, 67
overview, 65

P
Palla Ravi, 117
program trace, execution record, machine
        learning, surrogate fitness, behavioral
        diversity, implicit fitness sharing,
        lexicase selection, 170
Pseudoknots, 22

R
real-world application, 117
Replace Space With Newline problem, 191
Replace Space With Newline problem,
        Syllables problem, String Lengths
        Backwards problem, Negative To Zero
        problem, Double Letters problem,
        Scrabble Score problem, Checksum
        problem, Count Odds problem, 155
Riolo Rick, xiv

RNA, 22
rubrics, 67

S
Sequence-Structure Motifs, 23
Shahrzad Hormoz, 79
Silva Sara, 219
Simulation-based Optimization, 241
Sipper Moshe, 22
Spector Lee, 151
Stijven Sean, 241
Swan Jerry, 170
Symbolic Regression, 241
Symbolic Regression, Complexity Measures,
        Multi-objective Optimization,
        NSGA-II, 1

T
target-problems, 71
tdd-as-if-you-meant-it, 62
the-facilitator, 70
the-systems-turn, 72
the-tableau, 66
the-users-turn, 69
tournament selection, 185, 195
Trujillo Leonardo, 219
Truscott Philip, 203

V
Vaks Pavel, 22
Vanneschi Leonardo, 219
Veloso de Melo Vinícius, 39
Vladislavleva Ekaterina, 241

W
warrants, 69
why-a-warrant, 72
why-an-excuse, 60
Willem Lander, 241
Winkler Stephan M, 1
Worzel Bill, viii, xiv
Worzel W. P., 137

Y
Yousuf Aisha, 117

Z
Ziv-Ukelson Michal, 22

Printed in the United States
By Bookmasters